From Vehicles to Grid to Electric Vehicles to Green Grid

Many a Little
Makes a Miracle

From Vehicles to Grid to Electric Vehicles to Green Grid

Many a Little Makes a Miracle

Fuhuo Li
Henan Suda Electric Vehicle Technology Co. Ltd., China

Shigeru Kanemitsu
Henan Suda Electric Vehicle Technology Co. Ltd., China
& Kyushu Institute of Technology, Japan

Jianjie Zhang
Ministry of Industry and Information Technology
Equipment Industry Development Center, China

World Scientific

NEW JERSEY • LONDON • SINGAPORE • BEIJING • SHANGHAI • HONG KONG • TAIPEI • CHENNAI • TOKYO

Published by

World Scientific Publishing Co. Pte. Ltd.
5 Toh Tuck Link, Singapore 596224
USA office: 27 Warren Street, Suite 401-402, Hackensack, NJ 07601
UK office: 57 Shelton Street, Covent Garden, London WC2H 9HE

Library of Congress Control Number: https://lccn.loc.gov/2019024390

British Library Cataloguing-in-Publication Data
A catalogue record for this book is available from the British Library.

FROM VEHICLES TO GRID TO ELECTRIC VEHICLES TO GREEN GRID
Many a Little Makes a Miracle

ISBN 978-981-120-696-2

For any available supplementary material, please visit
https://www.worldscientific.com/worldscibooks/10.1142/11463#t=suppl

Preface

The 19th century physics classified substances into two: **matter** and **energy**, while the 20th century physics classified them into **bosons** and **fermions** with a new interpretation of energy as a form of matter. Fermions represent matter (including energy) and bosons, information (or the relation between matter). As a coincidence or a necessary consequence, this corresponds to the shift of IT (Information Technology) to ICT (Information Communication Technology), where information and communication in turn correspond to fermions and bosons, respectively. It is clear from our own experience that there occurs a cumulative error after a long use of ICT systems. Here the subtitle applies in its negative sense that many a little makes a mickle and the rather stereotyped notion of **information entropy** finds its proper place as an indicator of the increase of chaotic disorder. As is well-known, the terminology **entropy** (en for energy, trope for transformation) was introduced as an analogy to the *thermodynamic entropy* which plays an important role in thermodynamical systems. The notion of entropy applies to any system to indicate the increase of its disorder.

In a similar vein, the governing paradigm of the last centuries was that of *reductionism*, which, in the language of molecular biology, reads "genotype determines phenotype", i.e. collecting local data leads to the whole. Toward the end of the 20th century it has been noticed that although all the phenomena are assembled manifestations of some transformation of the elements in a system, reductionism has its limit and the whole is not just an assembly of the parts but that it is to be understood as a whole. A typical example is the discovery that homo sapiens and chimpanzees are less than 3 percent different in their DNAs. This has been thought of as a great discovery, a triumph of reductionism. However, from the global point of view, such a fact does not reveal anything about the difference between their behaviors. Reductionism has implied sectionalism and we have been living in the *world of Babel*, in the sense that there is no communication possible among scientists in different disciplines. The new paradigm of the 21st century is to be a unification of diverse disciplines into a bigger framework by which scientists with different backgrounds can also participate in solving the grave global issues of the 21st century, global warming, etc., from a higher standpoint. For this it is necessary for the generalist to have sound basics in fundamental sciences and books are needed which give a bird's eye view of basic sciences so that the enthusiastic readers can master the working knowledge thereof. However, there are only books that combine two subjects in a loose connection, such as mathematical physics, physical chemistry, etc., which are meant for studying just two subjects without thriving expansion to other disciplines. As far as we know, only Carbone and Gromov [Carbone and Gromov (2001)] has a flavor of unification, expounding molecular biology through mathematics.

There are a few disciplines which have made a remarkable progress in the 20th century. They are animal behaviorology, paleoclimatology (archeoclimatology), brain science, fluid dynamics, genetics, molecular biology (bioinformatics), to name a few. These are the ones which could not have been developed without the aid of computers. In the dawn of such disciplines, the main work has been piling up the data to form a complete database. But once a basic database is prepared, the next stage is more challenging and outsiders can also join to attack the main issues. Here we quote an impressive passage: *It is a curious fact that radical conceptual changes in science and technology are often initiated by men who did not receive their initial professional training in that science and technology. The person who first approaches the problems of a discipline* **with the mature perspective of an outsider**, *rather than being indoctrinated previously with the established methods and attitudes of the discipline, can sometimes point to unorthodox, though remarkably simple, solutions to those problems.* For example, the emerging importance of electric vehicles, **EV**s in Green Grid is in the spectrum, where by **Green Grid** (GG) we mean the advanced Smart Grid in which essential part of the whole electric power is supplied by distributed generation, i.e. it is generated by **renewable energy** depending on solar, biogas, geothermal energy, etc. Note that hydro-, tidal-, and wind-generations are included as part of solar (photovoltaic) generation (tidal generation is due to the influence of the moon, though). The "green" means that we think of nukes as not clean nor green. For at any moment a nuclear reactor starts a runaway and then it is almost impossible to control it, releasing fatal radiation. The danger of radiation is that when our bodies are exposed to it, it acts on water in our bodies and hydroxyl radical $^{\bullet}HO$ is produced. The hydroxyl radical then damages our genes. Cancer therapy depends on this very principle that hydroxyl radical damages the genes of cancer cells to make them die. Cf. §5.7. **Go green, get clean.**

The gene analysis of mitochondria made it clear that we human beings are direct descendants of the group of our ancient ancestors who lived in East Africa and to which belonged the **Mitochondria Eve**. It is said it could have been *a small group of some 20,000 members which survived 143,000 years ago.* The group has developed into a bigger one which then got divided into smaller groups, moving to the whole world. Around that time there were Sinanthropus Pekinensis, Pithecanthropus erectus and later Homo sapiens neanderthalensis. All of them extinguished. If we keep our genes exposed to toxication, our species Homo sapiens sapiens will also extinguish. No one wants to have his descendent suffer from gene damage.

All living creatures are trying to cope and get along with their variable environment by interpreting the information which they get through their sensory organs. In the case of humans, with their highly developed intelligence, they try to understand and interpret the outer world. As will be discussed in §5.7.1, Homo sapiens sapiens is blessed with very high activity of scavengers, i.e. annihilators (detoxicators) of active oxygen spices, and so it is rather long-living, thus it can inherit the knowledge that the tribe to which an individual belongs has inherited from their ancestors. This is the evolutionary strategy that Homo sapiens sapiens has chosen.

The present book is a culmination of attempts of unifying diverse scientific disciplines into a bigger framework, describing the idea of **applied whole-ism** by way of examples and worked-out exercises. One of the prevailing examples is EVs connected to GG, EV2GG. This is because vehicles are based on important principles in basic sciences and future mobilities are possible in the form of **EV2GG**. In order to unify into a seamless whole of many subjects including physics, chemistry, control theory, information science, mathematics, environmental science, medical sciences, etc., there must be underlying basic principles. The *continuo* is the **uplifting principle**, i.e. whenever possible, we uplift a few different material in a more general setting by similarity or similar. Some chapters have their own principle which we will state below.

We hope to provide a plausible picture of the near-future in 40 or more years from now on as well as an enlightening account of the real state-of-the-art of various universal world-wide problems of global warming, environmental pollution and their possible remedies. As one can see, many of the problems which are treated by models satisfying differential equations are accommodated in the exponential phenomena and we try to exhibit its ubiquity.

The subtitle suggests the two-sides of the principle of **cumulative effect** to the effect that collecting many small entities will give rise to a bigger entity. The good side is that assembling a lot of small but good entities will eventually attain a big enough good effect. The bad side can be termed as entropy increasing principle to the effect that even a minor harm, if harmful effects are accumulated a lot, will lead to a very bad effect. There are many examples of the bad effect, like being exposed to toxins including radiation, pesticides, taken-in polluted stuff, etc. There are also good effects including assembling cogenerative electricity, a large amount of EVs as a pool of electricity in addition to mobilities.

We shall now give descriptions of each chapter for the readers' convenience. In the first three chapters, we present the basics of three dynamics, mechanics, thermodynamics and fluid dynamics. They work as a concise text for mechanics.

Chapter 1 gives a brief description of Newtonian mechanics and transition to quantum information via quantum mechanics in §1. Since a moving vehicle obeys the Newton's law of motion, we first provide Newtonian mechanics and illustrate by vehicle dynamics in §1.3. We uplift both vehicle dynamics and Coriolis force in a general framework.

Anticipating the grid-controlled driving of EVs in the future, we shall also include basics of qubits theory which will lead to quantum computing. This is a solution to possible cyberattacks, §7.2.1.

Chapter 2 provides the basics of thermodynamics by way of the working principle of car engines. Traditionally, car engines are inner combustion engines which are one of the greatest inventions of the human kind, so that they make use of the difference in pressures before and after the combustion, where naturally the Boyle-Charles law shows its effect. This aspect will be scrutinized in §2.1. Then the combustion being a chemical reaction, naturally one takes into account the combustion process itself, where this aspect will be studied in §2.2.1. We present examples of uplifting: the Boyle-Charles law to the van der Waals state equation, the Otto and Diesel cycles to the Sabathé cycle.

Chapter 3 provides foundations of fluid dynamics in anticipation of its application to the theory of lubrication, which is one of the most effective method of reducing the friction. One of the highlights is Theorem 3.12 asserting that the Cauchy equation of motion implies the Navier-Stokes equation, i.e. a version of Newton's equation of motion is transformed into the governing equation in fluid dynamics. We give the Navier-Stokes equation in its fullest generality, i.e. for compressible flows. In most textbooks, it is stated for incompressible flows. We also state the 2-dimensional flows in relation to complex analysis.

Chapter 4 is a compilation of papers [Li *et al.* (2015)] and [Li *et al.* (2018a)] and Chapter 6 of books [Li *et al.* (2012)] and [Li *et al.* (2017)], giving rudiments of control theory. It also includes the contents of [Li and Li (2019)] which discusses an application of PSO (Particle Swarm Optimization) method for charging batteries. This chapter is a sort of watershed of physics (Chapters 1–3) and chemistry (Chapter 5).

The state is an interface between the past and the present and we may view all controls and possible cyberattack assessment as a dynamical system. We hope to study these dynamical systems from the point of view of the *transitional state* [Rozenfel'd and Yahinson (1966)] in which the Mikusinski distributions are used as opposed to the Sato hyperfunction in §3.14. It is quite natural to incorporate the distributions or hyperfunctions wherever the delta-function or similar non-functions appear. Cf. e.g. [Chakraborty *et al.* (2009)].

Chapter 5 gives various aspects of chemistry as needed to take a firm grip of the basics in environmental protection as well as mitigating the harm of toxins taken in by human bodies. The penetrating principle is the **similarity** (or adaptation) principle to the effect that compounds similar structures and forms behaving in a rather similar way. As manifestations of this, we dwell on the menace of the mistakenly taken-in toxins in human bodies, radiation, pesticides, too much nutrition, etc. as well as the adaptation method of adsorbing arsenal species by oyster shells—industrial waste.

Chapter 6 is related to meteorology partially depending on the paper [Li (2017)] and discusses climate change, carbon circulation and global warming from the point of view of circulation. One can make a better prediction by comparing the records kept in the earth. We state a weather cycle conjecture of a multiple of $\sim$700, of Dansgaard-Oeschger and Nishioka. The former depends on the historical record of the climate found in Greenland's ice core while the latter depends on the year-rings of old trees—dendroclimatology.

Chapter 7, by unifying the notions developed in previous chapters, provides the description of EV2GG, i.e. grid controlled electric vehicles. Human beings need mobilities and until the natural fossil fuel is exhausted, there will be a mixture of various types of cars. Although gasoline-propelled cars are thought of as one of the main sources of environmental pollution and greenhouse gases, the new much severer regulations make them of much lower harm to the air. The high-price of EVs and lower credit of driving range will force traditional cars to co-exist.

Smart grid operations essentially depend on computer control: many monitoring meters are equipped so that the system should run in order, and there is a **self-healing** function being installed so that the system can predict any possible danger and can prevent or at least mitigate it when it occurs, giving an option for the system operator to manually handle the danger, when appropriate. In the future system, it is hoped that new analytical programs are used which can perceive any possible menace to the system and as soon as they perceive it, they take an automatic action to prevent it or in the worst case, if it occurs, liquidate the effect to its minimum. This ensures safety of GG than individual autonomous control of an EV. This necessitates the grid control of all EVs which will serve as moving smartphone, so to say. All communications will be done wirelessly.

Thus, in order to realize Green Grid, it is crucial to assure cyber security and this leads to quantum computing in §1.5.3, completing the Uroboros circle. Another crucial ingredient in GG is enhanced batteries which can be charged *at rather low voltage*, e.g. by the light of the setting sun and the participation of enormous amount of EVs in the grid as a **pool of electricity**.

The GG is to transform the setting sun into the rising sun, and the EV2GG is to raise the welfare of the human being.

In Chapter 8, we give a Laplace transform foundation of the sampling theorem through the Paley-Wiener theorem in addition to the Fourier analysis method. In Chapter 4 (control theory) and Chapter 7 (EV2GG), (wireless) communication is essential. For this the signal being digital, signal restoration is the main issue. The celebrated sampling theorem (Shannon 1949) being essential in frequency analysis and in particular, in signal transmission, guarantees the restoration of the digital signal by sampling. We shall give several proofs of the Paley-Wiener theorem.

Last but not least, we would like to express our hearty thanks to Dr. Jay Mehta for his devoted and efficient help in preparing the final version of the book. We thank Mr. Y. Kamata for preparing some figures. As always, the book has been completed thanks to kind and constant encouragement by Ms. Kwong Lai Fun to whom we are very grateful.

Contents

Chapter 1

From mechanics to quantum information

Abstract

Cars are indispensable and convenient tools in modern society and have been used without much trouble since their invention some 150 years ago, save for the environmental pollution they have been causing. Like many other modern conveniences, almost no attention is paid to its working principle and people receive only vague images from the propaganda that the companies provide. But the car engines are far more crucial for grasping the mechanism that the human being has ever created than other black boxes including computers. In this chapter, we provide some basics on mechanics through the example of vehicle dynamics and explain the transition to quantum mechanics.

1.1 Newtonian mechanics

Let

$$\boldsymbol{x} = \boldsymbol{x}(t) = \begin{pmatrix} x \\ y \\ z \end{pmatrix} = \begin{pmatrix} x(t) \\ y(t) \\ z(t) \end{pmatrix} \tag{1.1}$$

be the position vector of a particle which is a function in time t, i.e. all components are functions in t. Hence the velocity is

$$\boldsymbol{v} = \boldsymbol{v}(t) = \frac{\mathrm{d}}{\mathrm{d}t}\boldsymbol{x} = \begin{pmatrix} u \\ v \\ w \end{pmatrix}, \tag{1.2}$$

where we write

$$u = \frac{\mathrm{d}x}{\mathrm{d}t}, \quad v = \frac{\mathrm{d}y}{\mathrm{d}t}, \quad w = \frac{\mathrm{d}z}{\mathrm{d}t}. \tag{1.3}$$

In many cases, one defines the acceleration as the derivative of the velocity with respect to time, i.e.

$$\boldsymbol{a} = \boldsymbol{a}(t) = \frac{\mathrm{d}}{\mathrm{d}t}\boldsymbol{v} = \frac{\mathrm{d}^2}{\mathrm{d}t^2}\boldsymbol{x} = \begin{pmatrix} \frac{\mathrm{d}^2 x}{\mathrm{d}t^2} \\ \frac{\mathrm{d}^2 y}{\mathrm{d}t^2} \\ \frac{\mathrm{d}^2 z}{\mathrm{d}t^2} \end{pmatrix}. \tag{1.4}$$

Newton's three **laws of motion** are
1. The first law (law of inertia) states that if the net force (the vector sum of all forces acting on an object) is zero, then the velocity of the object is constant.

$$\sum \boldsymbol{F} = \boldsymbol{o} \Longleftrightarrow \frac{\mathrm{d}}{\mathrm{d}t}\boldsymbol{v} = \boldsymbol{o}. \tag{1.5}$$

2. The second law states that if a body is accelerating, then there is a force on it which is directly proportional to the acceleration:

$$m\boldsymbol{a} = \boldsymbol{F}, \tag{1.6}$$

where m indicates the mass and $\boldsymbol{F}$ is the sum of all external forces acting on the body.
3. The third law (action-reaction law) states that if one object A exerts a force $\boldsymbol{F}_A$ (action) on a second object B, then B simultaneously exerts a force $\boldsymbol{F}_B$ (reaction) on A and

$$\boldsymbol{F}_A = -\boldsymbol{F}_B. \tag{1.7}$$

Newton's second law of motion is a typical example of the 2nd order system. (Cf. [Grodins (1963)].)

1.1.1 *Examples of the second-order systems*

- Newton's second law of motion

$$M\frac{\mathrm{d}^2 y}{\mathrm{d}t^2} + R\frac{\mathrm{d}y}{\mathrm{d}t} + Ky = e(t) = F, \tag{1.8}$$

where M is the inertance of mass, R is the viscous resistance of the dashpot and K is the spring stiffness.
Introducing the new parameters

$$\omega_n = \sqrt{\frac{K}{M}} : \text{natural angular frequency}$$

$$\zeta = \frac{R}{\sqrt{2\frac{K}{M}}} : \text{damping ratio},$$

(1.8) becomes

$$\frac{1}{\omega_n^2}\frac{\mathrm{d}^2 y}{\mathrm{d}t^2} + \frac{2\zeta}{\omega_n}\frac{\mathrm{d}y}{\mathrm{d}t} + y = \frac{1}{K}F. \tag{1.9}$$

Another example will appear in Chapter 4.

- Electrical circuits
  The electric current $i = i(t)$ flowing in an electrical circuit which consists of four ingredients: electromotive-force $e = e(t)$, resistance R, coil L and condenser C satisfies

$$L\frac{\mathrm{d}^2 i}{\mathrm{d}t^2} + R\frac{\mathrm{d}i}{\mathrm{d}t} + \frac{1}{C}i = e'(t). \tag{1.10}$$

1.2 Transitional states

Our objective in this section is to pave a road to clarifying both **Coriolis force** and **vehicle dynamics** by Theorem 1.1 which describes a transition from one state to another and gives a relation between derivatives in the inertia system and rotating system.

We refine arguments given in [Pedlovsky (1987), pp. 14-21], [Kondo (1993), p. 40], [Suzuki *et al.* (1984), pp. 62-70], [Tamagawa-seiki (2002), pp. 25-26, 141-149] etc. It turns out that up to a certain point, calculations amount to the Leibniz rule for differentiation of the product of two functions. In vehicle dynamics, analysis of characteristics of dynamics is made using the characteristic equation of the system of differential equations or Laplace transform thereof. Since this system of DEs can be solved explicitly either by linear algebra, Laplace transform, symbolic calculus, etc., we give an explicit solution based on linear algebra in Appendix A. Then based on the form of the solutions, we analyze the relation between steering angles and the radius of steady-state rotation.

The following theorem is fundamental for both Coriolis force and vehicle dynamics in the sense that by taking $\boldsymbol{b}$ to be the position vector resp. velocity vector, (1.12) (=(1.59)) gives the velocity vector resp. acceleration vector, where the reference to different coordinate system is to be noticed.

Theorem 1.1. *Let $\boldsymbol{w}_1, \boldsymbol{w}_2, \boldsymbol{w}_3$ be fundamental unit vectors constituting the curvilinear orthogonal coordinate system. Let*

$$\boldsymbol{b} = b_1\boldsymbol{w}_1 + b_2\boldsymbol{w}_2 + b_3\boldsymbol{w}_3, \tag{1.11}$$

where

$$b_j = \boldsymbol{b} \cdot \boldsymbol{w}_j.$$

Then we have

$$\left(\frac{\mathrm{d}\boldsymbol{b}}{\mathrm{d}t}\right)_I = \sum_{i=1}^{3} \frac{\mathrm{d}b_i}{\mathrm{d}t}\boldsymbol{w}_i + \sum_{i=1}^{3} b_i \frac{\mathrm{d}\boldsymbol{w}_i}{\mathrm{d}t} = \left(\frac{\mathrm{d}\boldsymbol{b}}{\mathrm{d}t}\right)_R + \sum_{i=1}^{3} b_i \frac{\mathrm{d}\boldsymbol{w}_i}{\mathrm{d}t}, \tag{1.12}$$

where the suffix means that the velocity is observed by the rotating observer (R) *or by a non-rotating inertial observer* (I), *respectively.*

Proof. The proof follows from the Leibniz rule. □

Remark 1.1. To express the second term in closed form, we use the expression in Lemma 1.1 below:

$$\frac{\mathrm{d}\boldsymbol{w}_i}{\mathrm{d}t} = \boldsymbol{\Omega} \times \boldsymbol{w}_i.$$

Hence the 2nd term is

$$\sum_{i=1}^{3} b_i(\boldsymbol{\Omega} \times \boldsymbol{w}_i) = \boldsymbol{\Omega} \times \sum_{i=1}^{3} b_i \boldsymbol{w}_i = \boldsymbol{\Omega} \times \boldsymbol{b},$$

so that Theorem 1.1 amounts to Theorem 1.4.

Theorem 1.2. *We choose the orthogonal curvilinear coordinate system in Theorem 1.1 as follows:*

$$\boldsymbol{w}_1 = \begin{pmatrix} \cos\gamma \\ \sin\gamma \\ 0 \end{pmatrix}, \boldsymbol{w}_2 = \begin{pmatrix} -\sin\gamma \\ \cos\gamma \\ 0 \end{pmatrix}, \boldsymbol{w}_3 = \begin{pmatrix} 0 \\ 0 \\ 1 \end{pmatrix} \tag{1.13}$$

and we take

$$\Omega = \gamma' \boldsymbol{w}_3. \tag{1.14}$$

Then for coordinates (1.11), (1.12) *reads*

$$\left(\frac{\mathrm{d}\boldsymbol{b}}{\mathrm{d}t}\right)_I = (b_1' - b_2\gamma')\boldsymbol{w}_1 + (b_2' + b_1\gamma')\boldsymbol{w}_2 + b_3'\boldsymbol{w}_3, \tag{1.15}$$

and its differentiated form is

$$\left(\frac{\mathrm{d}^2\boldsymbol{b}}{\mathrm{d}t^2}\right)_I = (b_1'' - 2b_2'\gamma' - b_1\gamma'^2 - b_2\gamma'')\boldsymbol{w}_1 + (b_2'' + 2b_1'\gamma' - b_2\gamma'^2 + b_1\gamma'')\boldsymbol{w}_2 + b_3''\boldsymbol{w}_3. \tag{1.16}$$

In the special case of the rotation of constant angular velocity $\omega > 0$, $\gamma = \gamma(t) = \omega t$, *then* (1.15) *amounts to*

$$\left(\frac{\mathrm{d}\boldsymbol{b}}{\mathrm{d}t}\right)_I = (b_1' - \omega b_2)\boldsymbol{w}_1 + (b_2' + \omega b_1)\boldsymbol{w}_2 + b_3'\boldsymbol{w}_3, \tag{1.17}$$

where $\boldsymbol{\Omega} = \omega\boldsymbol{w}_3$. (1.16) *reads*

$$\left(\frac{\mathrm{d}^2\boldsymbol{b}}{\mathrm{d}t^2}\right)_I = (b_1'' - 2\omega b_2' - b_1\omega^2)\boldsymbol{w}_1 + (b_2'' + 2\omega b_1' - b_2\omega^2)\boldsymbol{w}_2 + b_3''\boldsymbol{w}_3. \tag{1.18}$$

Proof. To express (1.12) in the form of (1.15) we need to use the formula

$$\boldsymbol{w}_1' = \gamma' \begin{pmatrix} -\sin\gamma \\ \cos\gamma \\ 0 \end{pmatrix} = \gamma' \boldsymbol{w}_3 \times \boldsymbol{w}_1 = \gamma' \boldsymbol{w}_2, \tag{1.19}$$

$$\boldsymbol{w}_2' = \gamma' \begin{pmatrix} -\cos\gamma \\ -\sin\gamma \\ 0 \end{pmatrix} = \gamma' \boldsymbol{w}_3 \times \boldsymbol{w}_2 = -\gamma' \boldsymbol{w}_2.$$

To prove (1.16) we differentiate (1.15) thereby using (1.19). □

The above proof is a little cumbersome because we do not have a rule for transition–differentiation. The following theorem gives an explicit rule for differentiation.

Theorem 1.3. *Let*

$$A_\gamma = \begin{pmatrix} \cos\gamma & -\sin\gamma \\ \sin\gamma & \cos\gamma \end{pmatrix}. \tag{1.20}$$

Then the linear transformation $\boldsymbol{y} = A_\gamma \boldsymbol{x}$ *by* A_γ *means the rotation (in the positive direction, i.e. counterclockwise) w.r.t. the origin by* γ. *Suppose* $\gamma = \gamma(t)$ *is a function in time* t. *Then*

$$\frac{\mathrm{d}}{\mathrm{d}t} A_\gamma = \gamma' A_{\gamma - \frac{\pi}{2}} = \gamma' A_\gamma S, \tag{1.21}$$

where

$$S = \begin{pmatrix} 0 & -1 \\ 1 & 0 \end{pmatrix}. \tag{1.22}$$

We have

$$\frac{\mathrm{d}}{\mathrm{d}t}(A_\gamma \boldsymbol{x}) = A_\gamma \left(\boldsymbol{x}' + \gamma' S \boldsymbol{x}\right) \tag{1.23}$$

and

$$\begin{aligned} \frac{\mathrm{d}^2}{\mathrm{d}t^2}(A_\gamma \boldsymbol{x}) &= \gamma' A_\gamma \left(S\boldsymbol{x}' + \gamma' S^2 \boldsymbol{x}\right) + A_\gamma \left(\boldsymbol{x}'' + \gamma' S \boldsymbol{x}' + \gamma'' S \boldsymbol{x}\right) \\ &= A_\gamma \left(\boldsymbol{x}'' + 2\gamma' S \boldsymbol{x}' + \left(-\gamma'^2 + \gamma'' S\right) \boldsymbol{x}\right). \end{aligned} \tag{1.24}$$

In the case where $\gamma = \omega t$ *(1.23) and (1.24) read*

$$\frac{\mathrm{d}}{\mathrm{d}t}(A_\gamma \boldsymbol{x}) = A_\gamma \left(\boldsymbol{x}' + \omega S \boldsymbol{x}\right) \tag{1.25}$$

and

$$\frac{\mathrm{d}^2}{\mathrm{d}t^2}(A_\gamma \boldsymbol{x}) = A_\gamma \left(\boldsymbol{x}'' + 2\omega S \boldsymbol{x}' + \left(-\omega^2\right) \boldsymbol{x}\right) \tag{1.26}$$

respectively.

We note that $A_\gamma^{-1} = A_{-\gamma}$ and $A_{-\gamma}A_{\gamma-\frac{\pi}{2}} = A_{-\frac{\pi}{2}} = -S$, so that

$$A_{\gamma-\frac{\pi}{2}} = A_\gamma S. \tag{1.27}$$

By augmenting the matrix $A_\gamma^{(k)}$ $(k = 0, 1)$ as

$$\tilde{A}_\gamma^{(k)} = \begin{pmatrix} A_\gamma^{(k)} & \boldsymbol{o} \\ \boldsymbol{o} & 1 \end{pmatrix} = \begin{pmatrix} \cos^{(k)}\gamma & -\sin^{(k)}\gamma & 0 \\ \sin^{(k)}\gamma & \cos^{(k)}\gamma & 0 \\ 0 & 0 & 1 \end{pmatrix} \tag{1.28}$$

we may view Theorem 1.2 as a special case of Theorem 1.3. (1.23) corresponds to (1.15). Indeed, writing $\frac{\mathrm{d}^2}{\mathrm{d}t^2}(A_\gamma \boldsymbol{b}) = \boldsymbol{a}_I$, $\frac{\mathrm{d}}{\mathrm{d}t}\boldsymbol{b} = \boldsymbol{v}_R$ etc. for the acceleration, velocity in the inertial or rotating system, we may express (1.23) and (1.24) as

$$\boldsymbol{v}_I = A_\gamma \left(\boldsymbol{v}_R + \gamma' S \boldsymbol{b}\right) \tag{1.29}$$

and

$$\boldsymbol{a}_I = A_\gamma \left(\boldsymbol{a}_R + 2\gamma' S \boldsymbol{v}_R + \left(-\gamma'^2 + \gamma'' S\right) \boldsymbol{b}\right) \tag{1.30}$$

which amounts to (1.15) and (1.16), respectively.

For vehicle dynamics we specify $\gamma = \omega t$ and

$$\begin{pmatrix} v_x \\ v_y \end{pmatrix} = \boldsymbol{b} = \boldsymbol{v}_R = V \begin{pmatrix} \cos\beta \\ \sin\beta \end{pmatrix}, \tag{1.31}$$

where $V > 0$ is the magnitude of the velocity and $\beta = \beta(t)$ is a slip angle which is infinitesimal. Then (1.23) reads

$$\boldsymbol{a}_I = A_\gamma \left(\boldsymbol{a}_R + \gamma' S \boldsymbol{v}_R\right). \tag{1.32}$$

Since

$$\frac{\mathrm{d}}{\mathrm{d}t}\boldsymbol{v}_R = V\beta' \begin{pmatrix} -\sin\beta \\ \cos\beta \end{pmatrix}, \tag{1.33}$$

the second factor on the right of (1.32) is

$$\begin{aligned} \boldsymbol{a}_R + \omega S V \begin{pmatrix} 0 & -1 \\ 1 & 0 \end{pmatrix} \begin{pmatrix} \cos\beta \\ \sin\beta \end{pmatrix} &= V \begin{pmatrix} -(\beta' + \omega)\sin\beta \\ (\beta' + \omega)\cos\beta \end{pmatrix} \\ &= V \begin{pmatrix} -(\beta' + \omega)\beta \\ (\beta' + \omega) \end{pmatrix} + O(\beta^2) \end{aligned}$$

as $\beta \to 0$. Hence

$$\boldsymbol{a}_I = VA_\gamma \begin{pmatrix} -(\beta' + \omega)\beta \\ (\beta' + \omega) \end{pmatrix}. \tag{1.34}$$

(1.34) in the coordinates (1.13) reads

$$\boldsymbol{a}_I = -V\beta(\beta' + \omega)\boldsymbol{w}_1 + V(\beta' + \omega)\boldsymbol{w}_2 + a_3\boldsymbol{w}_3, \tag{1.35}$$

where a_3 is the third component of $\boldsymbol{a}_I$ and we omitted the error term. Since for a vehicle, we may restrict to the 2-dimensional movement, Newton's equation of motion reads

$$m\boldsymbol{a}_I = -mV\beta(\beta' + \omega)\boldsymbol{w}_1 + mV(\beta' + \omega)\boldsymbol{w}_2. \tag{1.36}$$

1.3 Vehicle dynamics

During a high speed cornering of a vehicle, as a result of centripetal force, there occurs a discrepancy angle β called the **slip angle** (or sideslip angle) of its actual direction of travel from that of **steering angle** θ (the direction towards which a rolling wheel is pointing). I.e., it is the angle of the vector sum of wheel forward velocity v_x and lateral velocity v_y in (1.31). While moving, there occurs a friction between tires and road surface and the component of frictions perpendicular to the direction of travel is called the **cornering force** (CF). The safe cornering is possible as long as the sum of driving force and inertia force remains within the friction circle, a circle within which the tires keep grip on the road surface, i.e. by the balance of the centripetal force and the sum of cornering forces of all tires.

A yaw rotation with angle γ is a movement around the yaw axis of a vehicle that changes its direction of motion.

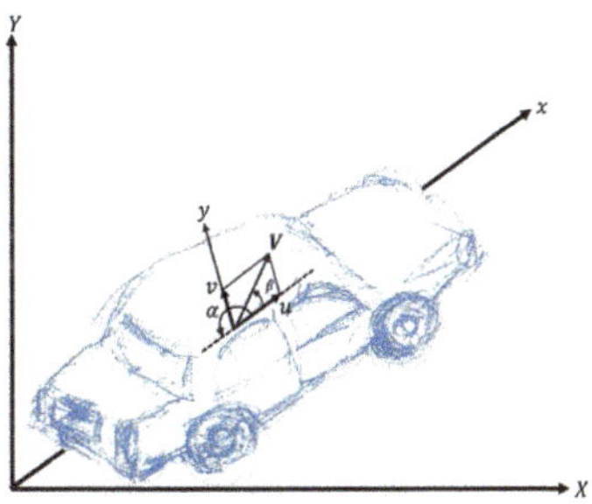

Fig. 1.1: Vehicle and coordinates

The results here are not only modifications but also refinements of those given in [EE (1999), pp. 24-28] and [Suzuki *et al.* (1984), pp. 62-70]. Recall

m	mass
V	velocity
I	yaw moment of inertia
γ	yaw angular velocity
$\beta = \beta_g$	sideslip angle
β_*	tire sideslip angle
θ_*	tire steering angle
l	wheelbase ($l = l_f + l_r$)
l_*	distance between axle and center of gravity
t_d	axle track ($d = 1/2t[d]$)
CF_j	cornering forces
K_*	cornering power
R	radius of rotation

Table 1.1: Vehicle parameters

Theorem 1.3 and note that $V \sin \beta \sim v\beta$ and that V is a constant. We use the following notation and parameters for a vehicle.

Here the suffix $*$ is either f or r as the case may be and the suffixes f and r refer to the object related to the front and rear tires. E.g. γ_f and β_r mean the front tire steering angle and rear tire slide slip angle respectively, while $\beta = \beta_g$ means the slip angle with respect to the center of gravity of the vehicle. For intended applications to in-wheel EV's, the distinction between front and rear tires will not be relevant and will be put equal. However, in order to include the case of 2WD front drive vehicles, we write some parameters with suffixes.

From the equilibrium of lateral forces we have

$$mv\left(\frac{\mathrm{d}\beta}{\mathrm{d}t} + \gamma\right) = m\left(\frac{\mathrm{d}}{\mathrm{d}t}(V\beta) + v\gamma\right) = m\boldsymbol{a} = \sum_{j=1}^{4} CF_j. \tag{1.37}$$

By geometric consideration we have

$$\beta_* = \theta_* - \left(\beta \pm \frac{l_*}{V}\gamma\right), \tag{1.38}$$

where the sign $\pm$ is to be $+$ in the case of in-wheel EVs. And the cornering force is given by

$$CF_* = K_*\beta_* = K_*\left(\beta + \frac{l_*}{V}\gamma\right) \tag{1.39}$$

with $K_* > 0$ indicating the cornering power.

(1.37) reads

$$mV\left(\frac{\mathrm{d}\beta}{\mathrm{d}t}+\gamma\right)=2(CF_f+CF_r)=2K_f\left(\beta+\frac{l_f}{V}\gamma\right)+2K_r\left(\beta+\frac{l_r}{V}\gamma\right). \tag{1.40}$$

The yawing angular momentum is given by

$$I_\gamma = I\frac{\mathrm{d}\gamma}{\mathrm{d}t}. \tag{1.41}$$

Hence the yawing equation of motion reads

$$I_\gamma = 2l_f CF_f - 2l_r CF_r. \tag{1.42}$$

Combining (1.40) and (1.42), we obtain

$$mV\frac{\mathrm{d}\beta}{\mathrm{d}t}+2(K_f+K_r)\beta+\left\{mV+\frac{2(l_fK_f-l_rK_r)}{V}\right\}\gamma=2(K_f\theta_f+K_r\theta_r) \tag{1.43}$$

$$I\frac{\mathrm{d}\gamma}{\mathrm{d}t}+2(l_fK_f-l_rK_r)\beta+\frac{2({l_f}^2K_f+{l_r}^2K_r)}{V}\gamma=2(l_fK_f\theta_f-l_rK_r\theta_r).$$

Suppose for simplicity that the steering angles be 0. Then (1.43) amounts to

$$\frac{\mathrm{d}}{\mathrm{d}t}\begin{pmatrix}\beta\\ \gamma\end{pmatrix}=-B\begin{pmatrix}\beta\\ \gamma\end{pmatrix} \tag{1.44}$$

where

$$B=\begin{pmatrix}\frac{2}{mV}(K_f+K_r) & 1+\frac{2(l_fK_f-l_rK_r)}{mV^2}\\ \frac{2}{I}(l_fK_f-l_rK_r) & \frac{2({l_f}^2K_f+{l_r}^2K_r)}{VI}\end{pmatrix}. \tag{1.45}$$

There are several methods known to solve (1.44). We shall give the method of linear algebra in Appendix A and below we give the basic data for the solution. The eigen-equation for $-B$ is

$$\begin{vmatrix}\lambda+\frac{2}{mV}(K_f+K_r) & 1+\frac{2(l_fK_f-l_rK_r)}{mV^2}\\ \frac{2}{I}(l_fK_f-l_rK_r) & \lambda+\frac{2({l_f}^2K_f+{l_r}^2K_r)}{VI}\end{vmatrix}=|\lambda E+B|=0. \tag{1.46}$$

Hence expanding we obtain the eigen-equation

$$\lambda^2+2b\lambda+c=0 \tag{1.47}$$

where

$$b=\frac{1}{mVI}\left(m\left({l_f}^2K_f+{l_r}^2K_r\right)+I\left(K_f+K_r\right)\right)>0 \tag{1.48}$$

and

$$c = \frac{4l - 2}{V^2} \frac{K_f K_r}{mI} (1 + \mathfrak{A} V^2) \tag{1.49}$$

and where $\mathfrak{A}$ is the **stability factor** defined by

$$\mathfrak{A} = -\frac{m}{2l^2} \frac{l_f K_f - l_r K_r}{K_f K_r}. \tag{1.50}$$

With steering angle and velocity constant, the vehicle makes a rotation with a fixed radius, called the steady-state rotation. In such a case, the derivatives of the yaw angle and the slip angle may be thought of 0 and (1.43) amounts to a linear equation with determinant given by

$$|B| = -2mV \left(l_f K_f - l_r K_r \right) + \frac{4}{V} K_f K_r l^2 = \frac{4}{V} K_f K_r l^2 \left(1 + \mathfrak{A} V^2 \right). \tag{1.51}$$

Solving the system we deduce that

$$\omega = \frac{V}{l} \frac{\theta_f - \theta_r}{(1 + \mathfrak{A} V^2)}, \tag{1.52}$$

whence the radius is

$$R = \frac{V}{\omega} = \left(1 + \mathfrak{A} V^2 \right) \frac{1}{\theta_f - \theta_r}. \tag{1.53}$$

Definition 1.1. From (1.53) we see that with steering angle $\theta_f - \theta_r$ the same and increasing the speed, the radius varies according to the sign of the quantity $I_f K_f - I_r K_r$ or stability factor $\mathfrak{A}$. If $\mathfrak{A} = 0$, the radius remains the same and the vehicle is said to have neutral steering property. If $\mathfrak{A} > 0$, then R increases as V increases, i.e. there is a deficit of steering angle to keep the radius of the original circle and the vehicle is said to have an **understeer** property. If $\mathfrak{A} < 0$, then the steering angle is excessive and the vehicle is said to have an **oversteer** property. In practice vehicles are designed to have weak understeer characteristic except for specified use.

Remark 1.2. Thus we may assume $\mathfrak{A} \geq 0$ and so $c > 0$ in (1.47) and it is often written in th form

$$\lambda^2 + 2\zeta\omega_n\lambda + \omega_n^2 = 0 \tag{1.54}$$

where

$$\zeta = \frac{1}{\omega_n} \frac{2}{mVI} \left(m \left(l_f{}^2 K_f + l_r{}^2 K_r \right) + I \left(K_f + K_r \right) \right) \tag{1.55}$$

with ω_n denoting the **natural angular frequency** defined by

$$\omega_n = \frac{2l}{V} \sqrt{\frac{K_f K_r}{mI} (1 + \mathfrak{A} V^2)}. \tag{1.56}$$

Hence vehicle dynamics is another instance of the second order system with decay cf. [Grodins (1963)], [Chakraborty *et al.* (2016), pp. 62-65] etc.

The solutions of (1.47) are $\lambda_1, \lambda_2 = -b \pm \sqrt{D}$, $D = b^2 - c$. The solutions of (1.44) are linear combinations of $e^{(-b+\sqrt{D})t}$ and $e^{(-b-\sqrt{D})t}$ or te^{-bt} and e^{-bt} according to whether the eigenvalues are different or the same.

In case $c > 0$ and $D \geq 0$, we have $D < b$ and the biggest term $e^{(-b+\sqrt{D})t}$ exponentially decreases.

In case $c > 0$ and $D < 0$, the fundamental solutions may be expressed as $e^{-bt} \cos \sqrt{-D}t$ and $e^{-bt} \sin \sqrt{-D}t$ so that the general solution is an oscillating function with exponential decay.

In case $c < 0$, we have $D > b$ and $e^{(-b+\sqrt{D})t}$ exponentially grows which is discarded as the case of oversteer.

The Laplace transform method can also be used for solving (1.44). Taking the Laplace transform of (1.43) and denoting $\beta(s) = L[\beta](s)$ and $\gamma(s) = L[\gamma](s)$ we obtain on the assumption that $\beta(0) = 0$

$$\begin{pmatrix} mVs + 2(K_f + K_r) & mV + \frac{2(l_f K_f - l_r K_r)}{V} \\ 2(l_f K_f - l_r K_r) & Is + \frac{2({l_f}^2 K_f + {l_r}^2 K_r)}{V} \end{pmatrix} \begin{pmatrix} \beta(s) \\ \gamma(s) \end{pmatrix} \tag{1.57}$$
$$= \begin{pmatrix} 2(K_f\theta_f + K_r\theta_r) \\ 2(l_f K_f \theta_f - l_r K_r \theta_r) \end{pmatrix}.$$

The determinant of the coefficient matrix in (1.57) may be expressed as

$$mVIs^2 + 2\left(m\left({l_f}^2 K_f + {l_r}^2 K_r\right) + I\left(K_f + K_r\right)\right)s - 2mV\left(l_f K_f - l_r K_r\right)$$
$$+ \frac{4}{V} K_f K_r l^2 \tag{1.58}$$

on writing

$$\left(K_f + K_r\right)\left({l_f}^2 K_f + {l_r}^2 K_r\right) = \left(l_f K_f - l_r K_r\right)^2 + K_f K_r (l_f + l_r)^2.$$

(1.58) amounts to the eigen-equation (1.47) and one may say that it is the characteristic equation for the system (1.47) of DE. Especially with $s = 0$, (1.58) gives the value of $|B|$ in (1.51).

There is an extensive exposition of the Laplace transform method in [Chakraborty *et al.* (2016)] using the partial fraction expansion and we omit further details.

1.4 Coriolis acceleration

In this section we state some basic facts about Coriolis acceleration (or Coriolis force) which is responsible for the global climate, esp. for the jet stream which flows in the stratosphere (8–13 km above the ground) at the speed of over 70 m/s. The Coriolis acceleration was discovered by

G. G. Coriolis in 1828 to the effect that a moving object in a rotating vessel is affected by the force whose direction is perpendicular to that of movement. Cf. [Pollack (1999)]. In the northern semi-sphere it tends to deflect the object to the right. The force is extremely weak: a man in a bullet train moving at 200 km receives the force of strength 20 g deflecting the movement to the right.

Theorem 1.1 in conjunction with Remark 1.1 reads

Theorem 1.4. *Let* $\boldsymbol{b} = \boldsymbol{b}(t)$ *be an arbitrary vector whose length is not necessarily constant and suppose that the coordinate system is rotating with angular velocity* $\boldsymbol{\Omega}$. *Then we have*

$$\left(\frac{\mathrm{d}\boldsymbol{b}}{\mathrm{d}t}\right)_I = \left(\frac{\mathrm{d}\boldsymbol{b}}{\mathrm{d}t}\right)_R + \boldsymbol{\Omega} \times \boldsymbol{b}, \tag{1.59}$$

where the suffix means that the velocity is observed by the rotating observer (R) *or by a non-rotating inertial observer* (I), *respectively.*

Lemma 1.1. *Let* $\boldsymbol{a} = \boldsymbol{a}(t)$ *denote an arbitrary vector of* **constant length** *which rotates with angular velocity* $\boldsymbol{\Omega}$. *Then we have*

$$\frac{\mathrm{d}}{\mathrm{d}t}\boldsymbol{a} = \boldsymbol{\Omega} \times \boldsymbol{a}. \tag{1.60}$$

Proof. Let α be the angle between $\boldsymbol{a}$ and $\boldsymbol{\Omega}$.

By the definition of the angular velocity, we have

$$\frac{\mathrm{d}\gamma}{\mathrm{d}t} = |\boldsymbol{\Omega}|. \tag{1.61}$$

From the definition of the rotation, the infinitesimal change $\frac{\mathrm{d}}{\mathrm{d}t}\boldsymbol{a}$ is perpendicular to $\boldsymbol{\Omega}$. It is also perpendicular to $\boldsymbol{a}$. For by assumption $0 = \frac{\mathrm{d}}{\mathrm{d}t}|\boldsymbol{a}|^2$ which is $\frac{\mathrm{d}}{\mathrm{d}t}\boldsymbol{a} \cdot \boldsymbol{a} = 2\frac{\mathrm{d}}{\mathrm{d}t}\boldsymbol{a} \cdot \boldsymbol{a}$ by the Leibniz rule. Hence

$$\boldsymbol{a} \perp \frac{\mathrm{d}}{\mathrm{d}t}\boldsymbol{a} = 0 \quad \text{or} \quad \frac{\mathrm{d}}{\mathrm{d}t}\boldsymbol{a} \perp \boldsymbol{a}.$$

Hence

$$\frac{\mathrm{d}}{\mathrm{d}t}\boldsymbol{a} // \boldsymbol{\Omega} \times \boldsymbol{a} \quad \text{or} \quad \frac{\mathrm{d}}{\mathrm{d}t}\boldsymbol{a} = c(\boldsymbol{\Omega} \times \boldsymbol{a}),$$

where $c > 0$ is a constant.

We determine c. By geometric consideration, the first approximation of

$$\Delta\boldsymbol{a} := \boldsymbol{a}(t + \Delta t) - \boldsymbol{a}(t)$$

is

$$|\boldsymbol{a}|(\sin\alpha)(\sin\Delta\gamma)\boldsymbol{n},$$

where

$$n = \frac{1}{|\boldsymbol{\Omega} \times \boldsymbol{a}|} \boldsymbol{\Omega} \times \boldsymbol{a}$$

is the normal vector. Recalling the Maclaurin expansion

$$\sin \gamma = \gamma + O\left(\gamma^3\right), \gamma \to 0, \tag{1.62}$$

we have

$$\Delta \boldsymbol{a} = |\boldsymbol{a}|(\sin \alpha)(\Delta\gamma)\boldsymbol{n} + O\left((\Delta\gamma)^3\right), \quad \Delta\gamma \to 0. \tag{1.63}$$

From (1.63) it follow that

$$\frac{\mathrm{d}}{\mathrm{d}t}\boldsymbol{a} = \lim_{\Delta t \to 0} \frac{\Delta \boldsymbol{a}}{\Delta t} = |\boldsymbol{a}| \sin \alpha \lim_{\Delta t \to 0} \frac{\Delta\gamma}{\Delta t}\boldsymbol{n} = |\boldsymbol{a}| \sin \alpha \frac{\mathrm{d}\gamma}{\mathrm{d}t}\boldsymbol{n} = |\boldsymbol{a}||\Omega| \sin \alpha \boldsymbol{n}, \tag{1.64}$$

by (1.61). Now in view of $|\boldsymbol{a}||\boldsymbol{\Omega}| \sin \alpha = |\boldsymbol{\Omega} \times \boldsymbol{a}|$, (1.64) reads

$$\frac{\mathrm{d}}{\mathrm{d}t}\boldsymbol{a} = |\boldsymbol{\Omega} \times \boldsymbol{a}|\boldsymbol{n},$$

which amounts to (1.60). □

Now let $\boldsymbol{r}$ be the rotation vector of an arbitrary fluid element. We first establish

Lemma 1.2. *The* **centripetal acceleration** $\boldsymbol{\Omega} \times (\boldsymbol{\Omega} \times \boldsymbol{r})$ *may be expressed as*

$$\boldsymbol{\Omega} \times (\boldsymbol{\Omega} \times \boldsymbol{r}) = -|\boldsymbol{\Omega}|^2 \boldsymbol{r}_\perp, \tag{1.65}$$

where $\boldsymbol{r}_\perp$ *is the distance vector, perpendicular to* $\boldsymbol{\Omega}$*, from* $\boldsymbol{\Omega}$ *to* $\boldsymbol{r}$*.*

Proof. By definition, we have

$$|\boldsymbol{\Omega} \times \boldsymbol{r}| = |\boldsymbol{\Omega} \times \boldsymbol{r}_\perp| = |\boldsymbol{\Omega}||\boldsymbol{r}_\perp| \sin \frac{\pi}{2} = |\boldsymbol{\Omega}||\boldsymbol{r}_\perp|.$$

Hence

$$\boldsymbol{\Omega} \times \boldsymbol{r} = \boldsymbol{\Omega} \times \boldsymbol{r}_\perp,$$

Using this in the triple vector product formula and noting that $\boldsymbol{\Omega} \cdot \boldsymbol{r}_\perp = 0$

$$\boldsymbol{a} \times (\boldsymbol{b} \times \boldsymbol{c}) = (\boldsymbol{a} \cdot \boldsymbol{c})\boldsymbol{b} - (\boldsymbol{a} \cdot \boldsymbol{b})\boldsymbol{c},$$

it follows that the left-hand side of (1.65) is $\boldsymbol{\Omega} \times (\boldsymbol{\Omega} \times \boldsymbol{r}_\perp) = -(\boldsymbol{\Omega} \cdot \boldsymbol{\Omega})\boldsymbol{r}_\perp$, which is the right-hand side of (1.65). □

Corollary 1.1. *A particle rotating along a circle of radius r with constant velocity $v = r\omega$, where ω indicates the angular velocity, receives the acceleration $r\omega^2$ which drags it to the right.*

Proof. We may apply Lemma 1.2 with $\gamma = \frac{\pi}{2}$, in which case $\boldsymbol{r} = \boldsymbol{r}_\perp$. □

Theorem 1.5. *Let $\boldsymbol{r}$ be the rotation vector with angular velocity $\boldsymbol{\Omega}$ of an arbitrary object with mass m. Then the momentum equation reads*

$$m\left(\frac{\mathrm{d}\boldsymbol{v}_R}{\mathrm{d}t}\right)_R = \mathcal{F} - 2m\boldsymbol{\Omega}\times\boldsymbol{v}_R - m\boldsymbol{\Omega}\times(\boldsymbol{\Omega}\times\boldsymbol{r}) - m\frac{\mathrm{d}\boldsymbol{\Omega}}{\mathrm{d}t}\times\boldsymbol{r}, \tag{1.66}$$

where $\boldsymbol{v} = \boldsymbol{v}_R$ is the vector observed in the rotating system, and the **Coriolis force** $-2\boldsymbol{\Omega}\times\boldsymbol{r}$ *appears as a force tending to deflect the moving fluid elements to the right.*

In the case of the constant angular velocity $\Omega = \begin{pmatrix}0\\0\\\omega\end{pmatrix}$ *we have*

$$m\left(\frac{\mathrm{d}\boldsymbol{v}_R}{\mathrm{d}t}\right)_R = \mathcal{F} - 2m\omega\begin{pmatrix}-v_\eta\\v_\xi\\0\end{pmatrix} - m\omega^2\begin{pmatrix}\xi\\\eta\\0\end{pmatrix}. \tag{1.67}$$

Proof. Applying Formula (1.59) to $\boldsymbol{r}$, we obtain

$$\boldsymbol{v}_I = \boldsymbol{v}_R + \boldsymbol{\Omega}\times\boldsymbol{r}, \tag{1.68}$$

where $\boldsymbol{v}_I = \left(\frac{\mathrm{d}\boldsymbol{v}}{\mathrm{d}t}\right)_I$ is the velocity observed in a non-rotating (inertial) system, while $\boldsymbol{v}_R = \left(\frac{\mathrm{d}\boldsymbol{r}}{\mathrm{d}t}\right)_R$ is the rotation velocity observed in a rotating system.

Applying (1.59) to $\boldsymbol{v}_I$, we obtain

$$\left(\frac{\mathrm{d}\boldsymbol{v}_I}{\mathrm{d}t}\right)_I = \left(\frac{\mathrm{d}\boldsymbol{v}_I}{\mathrm{d}t}\right)_R + \boldsymbol{\Omega}\times\boldsymbol{v}_I, \tag{1.69}$$

from which we want to eliminate $\left(\frac{\mathrm{d}\boldsymbol{v}_I}{\mathrm{d}t}\right)_R$. For this we differentiate (1.68) with respect to t to obtain

$$\begin{aligned}\left(\frac{\mathrm{d}\boldsymbol{v}_I}{\mathrm{d}t}\right)_R &= \left(\frac{\mathrm{d}\boldsymbol{v}_R}{\mathrm{d}t}\right)_R + \left(\frac{\mathrm{d}}{\mathrm{d}t}(\boldsymbol{\Omega}\times\boldsymbol{v})\right)\\ &= \left(\frac{\mathrm{d}\boldsymbol{v}_R}{\mathrm{d}t}\right)_R + \frac{\mathrm{d}\boldsymbol{\Omega}}{\mathrm{d}t}\times\boldsymbol{r} + \boldsymbol{\Omega}\times\left(\frac{\mathrm{d}\boldsymbol{r}}{\mathrm{d}t}\right)_R.\end{aligned}$$

Substituting this into (1.69) and using (1.68), we conclude that

$$\left(\frac{\mathrm{d}\boldsymbol{v}_I}{\mathrm{d}t}\right)_I = \left(\frac{\mathrm{d}\boldsymbol{v}_R}{\mathrm{d}t}\right)_R + \frac{\mathrm{d}\boldsymbol{\Omega}}{\mathrm{d}t}\times\boldsymbol{r} + \boldsymbol{\Omega}\times\boldsymbol{v}_R + \boldsymbol{\Omega}\times(\boldsymbol{v}_R + \boldsymbol{\Omega}\times\boldsymbol{r}),$$

whence

$$\left(\frac{\mathrm{d}\boldsymbol{v}_R}{\mathrm{d}t}\right)_R = \left(\frac{\mathrm{d}\boldsymbol{v}_I}{\mathrm{d}t}\right)_I - 2\boldsymbol{\Omega} \times \boldsymbol{v}_R - \boldsymbol{\Omega} \times (\boldsymbol{\Omega} \times \boldsymbol{r}) - \frac{\mathrm{d}\boldsymbol{\Omega}}{\mathrm{d}t} \times \boldsymbol{r}, \tag{1.70}$$

where $2\boldsymbol{\Omega} \times \boldsymbol{v}_R$ is the Coriolis acceleration and $\boldsymbol{\Omega} \times (\boldsymbol{\Omega} \times \boldsymbol{r})$ is the centripetal acceleration in Lemma 1.2.

In the case of the constant angular velocity $\Omega = \begin{pmatrix} 0 \\ 0 \\ \omega \end{pmatrix}$ we have

$$\boldsymbol{\Omega} \times \boldsymbol{r} = \omega \begin{pmatrix} -\eta \\ \xi \\ 0 \end{pmatrix} \tag{1.71}$$

and

$$\boldsymbol{\Omega} \times (\boldsymbol{\Omega} \times \boldsymbol{r}) = \omega^2 \begin{pmatrix} \xi \\ \eta \\ 0 \end{pmatrix}. \tag{1.72}$$

Hence (1.70) reads

$$\left(\frac{\mathrm{d}\boldsymbol{v}_R}{\mathrm{d}t}\right)_R = \left(\frac{\mathrm{d}\boldsymbol{v}_I}{\mathrm{d}t}\right)_I - 2\omega \begin{pmatrix} -v_\eta \\ v_\xi \\ 0 \end{pmatrix} - \omega^2 \begin{pmatrix} \xi \\ \eta \\ 0 \end{pmatrix}. \tag{1.73}$$

Hence, putting

$$\phi_c = \frac{1}{2} \mid \boldsymbol{\Omega} \mid^2 \mid \boldsymbol{r}_\perp \mid^2 = \frac{1}{2} \mid \boldsymbol{\Omega} \times \boldsymbol{r} \mid^2 .$$

We may write

$$\boldsymbol{\Omega} \times (\boldsymbol{\Omega} \times \boldsymbol{r}) = -\nabla \phi_c.$$

Hence the centripetal acceleration may be induced in the force rotated

$$\Phi = \phi + \phi_c.$$

Then, the momentum equation reads

$$\rho \left(\frac{\mathrm{d}\boldsymbol{v}}{\mathrm{d}t} + 2\boldsymbol{\Omega} \times \boldsymbol{v}\right) = -\nabla p + \rho \nabla \Phi + \mathcal{F}. \tag{1.74}$$

We have $\mathcal{F}(\boldsymbol{v}_I) = \mathcal{F}(\boldsymbol{v}_R)$ and (1.74) leads to (1.66). □

The Coriolis acceleration dominates for large scale flows or what amounts to the same thing, the **Rossby number**

$$\varepsilon = \frac{U}{2\Omega L}$$

is small (< 1).

Remark 1.3. We state (1.70) in the special circumstances of Theorem 1.2 with constant angular velocity. Since $\frac{d\Omega}{dt}$ vanishes, it reads

$$\left(\frac{d\boldsymbol{v}_R}{dt}\right)_R = \left(\frac{d\boldsymbol{v}_I}{dt}\right)_I - 2\boldsymbol{\Omega} \times \boldsymbol{v}_R - \boldsymbol{\Omega} \times (\boldsymbol{\Omega} \times \boldsymbol{r}). \tag{1.75}$$

Hence letting m be the mass of the rotating rigid body, we have the counterpart of (1.66):

$$m\frac{d\boldsymbol{v}}{dt} = -m|\boldsymbol{\Omega}|^2\boldsymbol{r}_\perp + \mathcal{F} - 2\boldsymbol{\Omega} \times \boldsymbol{r}. \tag{1.76}$$

1.5 From mechanics to quantum information

In this section we provide a transition from classical mechanics to quantum information via quantum mechanics. Quantum information developed in §1.5.3 will be an essential tool in the near future and in particular in EV participation in GG through wireless communications systems, cf. §7.2.1. We touch continuous variable quantum information a little and state mainly rudiments of the theory of qubits elucidating it as part of linear algebra. Being simple, qubits are more feasible for easier implementation through quantum gates.

For some basic facts about Hilbert spaces and their linear operators, cf. [Li *et al.* (2017), §5.1], where we called them additive operators. We shall use the notation in [Li *et al.* (2017), §5.1] rather than the commonly used one in quantum mechanics. A typical difference is that T^* means the conjugate operator, i.e.

$$(T\Psi_1, \Psi_2) = (\Psi_1, T^*\Psi_2) \tag{1.77}$$

rather than the complex conjugate $\bar{T}$ (which is denoted T^* in quantum mechanics where the symbol $T^\dagger$ is used to mean the conjugate operator).

In quantum mechanics, one assumes that the totality of all states of a system form a normed $\mathbb{C}$-vector space V and that all (quantum) mechanical quantities are expressed as Hermitian operators (or self-adjoint operators)

$T : V \to V$ satisfying $T^* = T$, cf. [Li *et al.* (2017), p. 177], where V is to be an L^2-space as seen below.

Taking for granted that every physical quantity T can be represented by a linear operator, which by abuse of language is denoted by the same letter, T, we shall see Hermitian operators are the ones that are needed. For the **expectation value** of T is given as the inner product

$$(\psi, T\psi) = \langle\psi|T\psi\rangle = \int \psi^* T\psi \, \mathrm{d}\boldsymbol{r}, \tag{1.78}$$

where $(T\psi, \psi)$ indicates the inner product in the vector space V and $\boldsymbol{r} = (x, y, z)$ is the position vector. Since expectation value is the average of the measured real values of a physical quantity, it must be real. For this it suffices that the conjugate operator T^* exists such that $T^* \supset T$, i.e. T be symmetric. Symmetric operators are not always self-adjoint but physical interpretation requires that operators which represent measurable physical quantities must be self-adjoint, it is customary to use symmetric and Hermitian synonymously.

For a Hermitian operator T, the eigenvectors $|\psi\rangle$ belonging to its eigenvalue $\lambda\,(\in \mathbb{R})$ are viewed as the quantum state whose mechanical quantity is equal to λ.

The Hermitian operator H expressing the total energy of a system is called the **Hamiltonian**. In most of quantum physics books, the Hamiltonian is denoted $\hat{H}$ but we use the same letter H as with the Hamiltonian function.

The wave function $|\Psi\rangle = |\Psi(\boldsymbol{r}, t)\rangle$, which is a function in the position vector and time t, is to satisfy the (time dependent) **Schrödinger equation**

$$i\hbar\frac{\partial}{\partial t}|\Psi(\boldsymbol{r}, t)\rangle = H|\Psi(\boldsymbol{r}, t)\rangle, \tag{1.79}$$

where $|\Psi(\boldsymbol{r}, t)\rangle$ indicates the ket vector to be introduced in §1.5.3,

$$\hbar = \frac{h}{2\pi} \tag{1.80}$$

is called the (reduced) **Planck constant** and where

$$h = 2\pi\hbar = 6.624 \times 10^{-27} \text{ erg} \times \text{sec}$$

is the Planck constant.

More concretely, consider a system of N point particles $\boldsymbol{r}_j = (x_j, y_j, z_j)$ with mass m_i, $1 \le j \le N$ moving around a three-dimensional space under the influence of forces given as a potential energy $V = V(\boldsymbol{R}, t)$, where

$$\boldsymbol{R} = (\boldsymbol{r}_1, \cdots, \boldsymbol{r}_N) = (x_1, \cdots, z_N) \in \mathbb{R}^{3N} \tag{1.81}$$

is a $3N$-dimensional vector. For this the Hamiltonian H in (1.79) reads

$$H = -\frac{\hbar^2}{2}\sum_{j=1}^{N}\frac{1}{m_j}\Delta_j + V, \tag{1.82}$$

where Δ_j is the Laplacian in (3.13) operating on each $\boldsymbol{r}_i$ and the Schrödinger equation (1.79) amounts to the non-relativistic Schrödinger equation in position basis

$$i\hbar\frac{\partial}{\partial t}\Psi(\boldsymbol{R}, t) = H\Psi(\boldsymbol{R}, t), \tag{1.83}$$

where $\Psi(\boldsymbol{R}, t)$, called the (position-space) wave function, describes the state of the system at time t. In this case, there is a postulate (M. Born) that the probability of the positions of N particles lying in a certain region $D \subset \mathbb{R}^{3N}$ is to be proportional to

$$\int \cdots \int_D |\Psi(\boldsymbol{R}, t)|^2 \,\mathrm{d}\boldsymbol{R}. \tag{1.84}$$

For this we must have the probability of the total event to be 1, so that

$$\int \cdots \int_{\mathbb{R}^{3N}} |\Psi(\boldsymbol{R}, t)|^2 \,\mathrm{d}\boldsymbol{R} < \infty, \tag{1.85}$$

i.e. the L^2-space. $L^2(\mathbb{R}^{3N})$ is the place where the wave functions are to lie.

In the form (1.82), we may easily check the condition for H to be symmetric. Let $D \subset \mathbb{R}^{3N}$ be either bounded or equal to the whole set. In the former case, we consider only those wave functions which are 0-extensions, i.e. $\Psi(\boldsymbol{R}, t) - 0$ if $\boldsymbol{R} \notin D$ and in the case $D = \mathbb{R}^{3N}$, we take account that $\Psi(\boldsymbol{R}, t) \to 0$ quickly enough as one of the position variables tends to ∞. We refer to this as a boundary condition. It suffices to consider the case $\boldsymbol{R}$ x, say. By integration by parts, we have ($\boldsymbol{R}'$ denoting the $(3N-1)$-dimensional vector)

$$\begin{aligned}
&\int \cdots \int_D \overline{\Psi_1(\boldsymbol{R}, t)}\frac{\partial^2}{\partial^2 x}\Psi_2(\cdots, x, \cdots, t)\,\mathrm{d}\boldsymbol{R} \\
&= -\int \cdots \int_{\partial D} \frac{\partial}{\partial x}\overline{\Psi_1(\boldsymbol{R}, t)}\frac{\partial}{\partial x}\Psi_2(\cdots, x, \cdots, t)\,\mathrm{d}\boldsymbol{R}' \\
&= \int \cdots \int_D \frac{\partial^2}{\partial^2 x}\overline{\Psi_1(\boldsymbol{R}, t)}\Psi_2(\cdots, x, \cdots, t)\,\mathrm{d}\boldsymbol{R},
\end{aligned}$$

or (1.77) with $T^* = T = H$.

We shall explain how the Hamiltonian (1.82) is deduced from the Hamiltonian function. For a conservative system (in which the potential energy does not depend on time) the Hamiltonian function $H = H(\boldsymbol{R})$ is the total energy (kinetic and potential) of the system:

$$H = H(\boldsymbol{R}) = T(\boldsymbol{p}) + V(\boldsymbol{R}) = W. \tag{1.86}$$

Here the momentum p_k related to the direction x_k is $m_k\dot{x}_k$, so that

$$\boldsymbol{p} = (p_1, \cdots, p_N) = (m_1\dot{x}_1, \cdots, m_N\dot{x}_N) \tag{1.87}$$

and the **kinetic energy** $T = T(\boldsymbol{p})$ is

$$T = \frac{1}{2}\sum_{k=1}^{n} m_k(\dot{x}_k^2 + \dot{y}_k^2 + \dot{z}_k^2) = \frac{1}{2}\sum_{k=1}^{n} \frac{1}{m_k}p_k^2. \tag{1.88}$$

Substituting (1.88) into (1.86), we deduce that

$$H(\boldsymbol{R}) = \frac{1}{2}\sum_{k=1}^{n} \frac{1}{m_k}p_k^2 + V(\boldsymbol{R}) = W. \tag{1.89}$$

Making the **replacement**

$$p_k \leftrightarrow \frac{h}{2\pi i}\frac{\partial}{\partial x_k} = -i\hbar\frac{\partial}{\partial x_k}, \quad W \leftrightarrow -\frac{h}{2\pi i}\frac{\partial}{\partial t} = i\hbar\frac{\partial}{\partial t}, \tag{1.90}$$

we obtain the Hamiltonian (1.82).

This operates on the wave function $\Psi(\boldsymbol{R}, t)$ as

$$H\Psi(\boldsymbol{R}, t) = -\frac{\hbar^2}{2}\sum_{j=1}^{N} \frac{1}{m_j}\Delta_j\Psi(\boldsymbol{R}, t) + V\Psi(\boldsymbol{R}, t) = i\hbar\frac{\partial\Psi}{\partial t}. \tag{1.91}$$

Hence the wave equation amounts to (1.83).

In case the boundary conditions can be so chosen that the Hamiltonian is Hermitian, then (1.91) can be solved using the spectral decomposition. [Yoshida (1956), pp. 172-185]:

$$\Psi(\boldsymbol{R}, t) = U_t\Psi(\boldsymbol{R}, 0), \quad U_t = e^{-\frac{i}{\hbar}Ht} \tag{1.92}$$

where e^B is the exponential function in the bounded operator. Hence this may be thought of as a manifestation of the exponential phenomenon. We note that U_t is a **unitary operator** [Li *et al.* (2017), p. 175]. We add a standard characterization [Yoshida (1956), p. 41].

Theorem 1.6. *For a bounded operator U to be unitary it is necessary and sufficient that*

$$U^* = U^{-1}. \tag{1.93}$$

Now we try to solve (1.91) under the assumption that Ψ is of variable separable type

$$\Psi(\boldsymbol{R}, t) = \varphi(t)\psi(\boldsymbol{R}). \tag{1.94}$$

Substituting this in (1.91), we have

$$\left(-\frac{\hbar^2}{2}\sum_{j=1}^{N}\frac{1}{m_j}\Delta_j\psi(\boldsymbol{R}) + V(\boldsymbol{R})\psi(x)\right)\varphi(t) = i\hbar\psi(\boldsymbol{R})\frac{\mathrm{d}\varphi(t)}{\mathrm{d}t}, \tag{1.95}$$

whence dividing both sides by $\varphi(t)\psi(\boldsymbol{R})$ we have

$$\frac{1}{\psi(\boldsymbol{R})}\left(-\frac{\hbar^2}{2m}\frac{\partial^2\psi}{\partial x^2} + V(\boldsymbol{R})\psi(\boldsymbol{R})\right) = \frac{1}{\varphi(t)}\left(i\hbar\frac{\mathrm{d}\varphi(t)}{\mathrm{d}t}\right). \tag{1.96}$$

Since the left-hand side of (1.96) is a function in $\boldsymbol{R}$ only and the right-hand side in t only, it follows that (1.96) must be a constant, say W. Hence (1.96) is equivalent to the system

$$\begin{cases} \frac{\mathrm{d}\varphi(t)}{\mathrm{d}t} & = -\frac{i}{\hbar}W\varphi(t) \\ -\frac{\hbar^2}{2}\sum_{j=1}^{N}\frac{1}{m_j}\Delta_j\psi(\boldsymbol{R}) + V(\boldsymbol{R})\psi(\boldsymbol{R}) & = W\psi(\boldsymbol{R}). \end{cases} \tag{1.97}$$

The second equation in (1.97) is often written as

$$\sum_{j=1}^{N}\frac{1}{m_j}\Delta_j\psi(\boldsymbol{R}) + \frac{2}{\hbar^2}(W - V)\psi(\boldsymbol{R}) = 0 \tag{1.98}$$

and called the Schrödinger amplitude equation for a conservative system of point particles. $W = W_n$ represents the energy of the system at its various stationary states and is called the **characteristic energy values** or **eigenvalues of the wave equation**.

Since the first equation in (1.97) is one for the first-order reaction, the solution is given by $e^{-\frac{i}{\hbar}W_n t}$ by Proposition 1.1. Hence by the principle of superposition, the general solution $\Psi(x, t)$ of (1.91) is given by

$$\Psi(\boldsymbol{R}, t) = \sum_n a_n\Psi_n(\boldsymbol{R}, t) = \sum_n a_n\psi_n(\boldsymbol{R})e^{-i\frac{W_n}{\hbar}t}. \tag{1.99}$$

1.5.1 *WKB method*

The Wentzel-Kramers-Billouin Method (WKB method) consists in giving an approximation to the solution of the wave equation. The first term leads to the result obtained by classical mechanics (Newtonian mechanics), the second term to the old-quantum theory result, and higher terms to corrections which brings in the effects characteristic of the new (quantum) mechanics.

(1.98) reads for $\boldsymbol{R} = x$

$$\frac{\mathrm{d}^2\psi}{\mathrm{d}x^2} + \frac{2m}{\hbar^2}(W - V)\psi(x) = 0. \tag{1.100}$$

Since this is a linear DE of degree 2 with variable coefficients, we apply the **Lagrange constant variation method** to find its solutions. Let

$$\psi = \psi(x) = e^{q(x)}, \tag{1.101}$$

say, where

$$q(x) = \frac{i}{\hbar}\int y \,\mathrm{d}x \tag{1.102}$$

and $y = y(x)$. Then

$$\frac{\mathrm{d}\psi}{\mathrm{d}x} = \frac{i}{\hbar} y e^{q(x)}. \tag{1.103}$$

Hence differentiating again, we obtain

$$\frac{\mathrm{d}^2\psi}{\mathrm{d}x^2} = \frac{i}{\hbar}\left(\frac{\mathrm{d}y}{\mathrm{d}x} + \frac{i}{\hbar}y^2\right) e^{q(x)}. \tag{1.104}$$

Substituting (1.104) and (1.101) into (1.100) and dividing both sides by $\frac{1}{\hbar^2}e^{q(x)}$, we transform (1.100) into

$$-\frac{\hbar}{i}\frac{\mathrm{d}y}{\mathrm{d}x} + y^2 + 2m(W - V) = 0$$

or

$$\frac{\hbar}{i}\frac{\mathrm{d}y}{\mathrm{d}x} = 2m(W - V) - y^2 = p^2 - y^2, \tag{1.105}$$

say, where

$$p = \sqrt{2m(W - V)} \tag{1.106}$$

is the classical expression for the momentum.

In [Pauling and Wilson (1935)], they *a priori* view y as a function in h and expand it into the Maclaurin series in $t := \frac{\hbar}{i}$:

$$y = y(x, h) = \sum_{n=0}^{\infty} y_n t^n, \tag{1.107}$$

where y_j are functions in x. We compare the coefficients of (1.107) and (1.105) in the form

$$t\frac{\mathrm{d}y}{\mathrm{d}x} = t\left(\sum_{n\ 0}^{\infty} y_n' t^n\right) = p^2 - y^2 = p^2 - (y_0^2 + 2y_0y_1t + (y_1^2 + 2y_0y_2)t^2 + \cdots). \tag{1.108}$$

Hence

$$p^2 - y_0^2 = 0, \quad 2y_0y_1 = -y_0', \quad y_1^2 + 2y_0y_2 = -y_1', \cdots \tag{1.109}$$

so that

$$y_0 = p, \quad y_1 = \frac{-y_0'}{2y_0} = \frac{-p'}{2p}, \quad y_2 = -\frac{1}{2y_0}\left(y_1' + y_1^2\right)\cdots. \tag{1.110}$$

Since

$$p' = \frac{-2mV'}{2p}, \tag{1.111}$$

it follows that

$$y_1 = \frac{2mV'}{(2p)^2} = \frac{V'}{4(W-V)}. \tag{1.112}$$

Substituting the expressions for y_1' and y_1^2 in the third equality in (1.110), we find that

$$\begin{aligned} y_2 &= -\frac{1}{2p}\frac{1}{16(W-V)^2}(4V''(W-V)+5V'^2) \\ &= -\frac{1}{32}\frac{1}{\sqrt{2m}(W-V)^{5/2}}(4V''(W-V)+5V'^2). \end{aligned} \tag{1.113}$$

Incorporating the coefficients y_0, y_1, y_2 from (1.107), (1.112) and (1.113), we have an approximation to (1.102)

$$\begin{aligned} y &= y_0 + \frac{\hbar}{i}y_1 + \left(\frac{\hbar}{i}\right)^2 y_2 + \cdots = p + \frac{\hbar}{i}\frac{V'}{4(W-V)} \\ &\quad - \left(\frac{\hbar}{i}\right)^2 \frac{1}{32}\frac{1}{\sqrt{2m}(W-V)^{5/2}}(4V''(W-V)+5V'^2) + \cdots. \end{aligned} \tag{1.114}$$

Substituting (1.114) up to the second term and noting that by (1.112)

$$\int y_1 \,\mathrm{d}x = -\frac{1}{4}\log(W-V) + C, \tag{1.115}$$

we find an approximation

$$\psi = M(W-V)^{-\frac{1}{4}} e^{\int \sqrt{2m(W-V)}\,\mathrm{d}x} \tag{1.116}$$

with M a constant. The probability distribution function is

$$\bar{\psi}\psi \approx C(W-V)^{-\frac{1}{2}} \approx \frac{1}{p} \tag{1.117}$$

in conformity with the classical result that the probability of finding a particle in a range $\mathrm{d}x$ is inversely proportional to the velocity in that range and the velocity is proportional to p.

Quantization occurs when one attempts to extend the region for W by an associated condition

$$\int_C y\,\mathrm{d}x = 2\pi h n, \quad m = 0, 1, 2, \cdots, \tag{1.118}$$

where the integral is a certain contour integral. The first approximation in (1.114) amounts to the classical quantum theory condition (1.118) with p in place of y. The second approximation in (1.114) gives

$$\int_C p\,\mathrm{d}x = 2\pi h\left(n + \frac{1}{2}\right), \quad m = 0, 1, 2, \cdots. \tag{1.119}$$

1.5.2 *Eigenfunctions*

Example 1.1. We consider the case where the quantum state $\mathbf{\Psi} = \mathbf{\Psi}(t)$ varies with time variable t only according to the Schrödunger equation (1.79), which appears in the study of molecular orbitals. Cf. [Chakraborty *et al.* (2009), pp. 16-22].

If $H\mathbf{\Psi}(t) = E\mathbf{\Psi}(t)$ and the eigenvalue E is the **energy level** of the system, cf. the first equation in (1.97). The solution of this equation is called the stationary state on the ground where its expectation does not change with time. The energy level means the values of the energy which the stationary state can assume. (1.79) amounts to

$$\frac{\mathrm{d}}{\mathrm{d}t}\mathbf{\Psi}(t) = \frac{-iE}{\hbar}\mathbf{\Psi}(t) \tag{1.120}$$

and the solution given by (1.92) amounts to

$$\mathbf{\Psi}(t) = e^{-\frac{iEt}{\hbar}}\mathbf{\Psi}(0). \tag{1.121}$$

Example 1.2. Writing in (1.100)

$$\alpha = \frac{1}{\hbar^2}p^2, \tag{1.122}$$

where p is defined by (1.106), we find that (1.100) amounts to the homogeneous equation

$$y'' + \alpha y = 0. \tag{1.123}$$

In the case where α is a constant (we may suppose $\alpha > 0$), this can be solved e.g. by the method of diagonalization, cf. Theorem 1.12.

Here we present one more effective method—the Laplace transform method, cf. e.g. [Kanemitsu and Tsukada (2007), pp. 154-161]. We obtain

$$Y(s) = L[y](s) = \int_0^\infty e^{-st}y(t)\,\mathrm{d}t = \frac{y(0)s + y'(0)}{s^2+\alpha} \tag{1.124}$$

By partial fraction expansion,

$$Y(s)=L[y](s)=\frac{y(0)}{2}\left(\frac{1}{s-i\sqrt{\alpha}}+\frac{1}{s+i\sqrt{\alpha}}\right)+\frac{y'(0)}{2i\sqrt{\alpha}}\left(\frac{1}{s-i\sqrt{\alpha}}-\frac{1}{s+i\sqrt{\alpha}}\right). \tag{1.125}$$

Hence by the Laplace inversion, the solution to (1.123) is

$$y(x) = y(0)\cos\sqrt{\alpha}x + \frac{y'(0)}{\sqrt{\alpha}}\sin\sqrt{\alpha}x = A\sin(\sqrt{\alpha}x+\theta), \tag{1.126}$$

say.

For curiosity we state a more general case.

Exercise 1. Given two quadratic polynomials $f_j(s), j = 1,2$

$$\mathbb{R}[s] \ni f_j(s) = s^2 + a_j s + b_j = (s-\lambda_j)(s-\bar{\lambda}_j), \quad \lambda_j = \frac{-a_j+\sqrt{D_j}i}{2}, \tag{1.127}$$

$$D_j = 4b_j - a_j^2 > 0, \quad \lambda_1 \neq \lambda_2,$$

find the coefficients in the partial fraction expansion by residue calculus

$$f(s) = \frac{1}{f_1(s)f_2(s)} = \frac{A_1}{s-\lambda_1} + \frac{\bar{A}_1}{s-\bar{\lambda}_1} + \frac{A_2}{s-\lambda_2} + \frac{\bar{A}_2}{s-\bar{\lambda}_2}. \tag{1.128}$$

Solution.

$$A_1 = \lim_{s\to\lambda_1}(s-\lambda_1)f(s) = \lim_{s\to\lambda_1}\frac{1}{(s-\bar{\lambda}_1)f_2(s)} = \frac{1}{(\lambda_1-\bar{\lambda}_1)f_2(\lambda_1)} \tag{1.129}$$

$$= \frac{-i}{\sqrt{D_1}}\frac{f_2(\bar{\lambda}_1)}{|f_2(\lambda_1)|^2} = \frac{-2i}{\sqrt{D_1}}\frac{(a_1-a_2)a_1-2b_1+2b_2+(a_1-a_2)\sqrt{D_1}i}{((a_1-a_2)a_1-2b_1+2b_2-(a_1-a_2)^2+(a_1-a_2)^2D_1}$$

$$= \frac{2}{\sqrt{D_1}}\frac{(a_1-a_2)\sqrt{D_1}-i\left((a_1-a_2)a_1-2b_1+2b_2\right)}{((a_1-a_2)a_1-2b_1+2b_2)^2+(a_1-a_2)^2D_1}$$

since

$$f_2(\lambda_1) = \frac{1}{2}\left((a_1 - a_2)a_1 - 2b_1 + 2b_2 - (a_1 - a_2)\sqrt{D_1}i\right). \tag{1.130}$$

Changing the suffixes, we have

$$A_2 = \frac{2}{\sqrt{D_2}} \frac{(a_2 - a_1)\sqrt{D_2} - i\left((a_2 - a_1)a_2 - 2b_2 + 2b_1\right)}{\left((a_2 - a_1)a_2 - 2b_2 + 2b_1\right)^2 + (a_2 - a_1)^2 D_2}. \tag{1.131}$$

Exercise 2. With A_j in (1.129), (1.131) find

$$2\,\mathrm{Re}\left(A_j e^{\lambda_i i t}\right). \tag{1.132}$$

Solution. By Euler's identity

$$2\,\mathrm{Re}\left(A_j e^{\lambda_i} it\right) = 2e^{-\frac{a_j}{2}}\left((\mathrm{Re}\,A_j)\cos\frac{\sqrt{D_j}}{2}t - (\mathrm{Im}\,A_j)\sin\frac{\sqrt{D_j}}{2}t\right) \tag{1.133}$$

$$= 2e^{-\frac{a_j}{2}}\left(\frac{2}{\sqrt{D_1}}\frac{(a_1 - a_2)\sqrt{D_1}}{\left((a_1 - a_2)a_1 - 2b_1 + 2b_2\right)^2 + (a_1 - a_2)^2 D_1}\cos\frac{\sqrt{D_j}}{2}t\right.$$

$$\left. + \frac{2}{\sqrt{D_1}}\frac{(a_1 - a_2)a_1 - 2b_1 + 2b_2}{\left((a_1 - a_2)a_1 - 2b_1 + 2b_2\right)^2 + (a_1 - a_2)^2 D_1}\sin\frac{\sqrt{D_j}}{2}t\right).$$

Exercise 3. In parallel to Exercise 1 prove that the coefficients in the partial fraction expansion

$$f(s) = \frac{s - a}{f_1(s)f_2(s)} = \frac{A_1}{s - \lambda_1} + \frac{\bar{A}_1}{s - \bar{\lambda}_1} + \frac{A_2}{s - \lambda_2} + \frac{\bar{A}_2}{s - \bar{\lambda}_2} \tag{1.134}$$

are given by

$$A_1 = \frac{\lambda_1 - a}{(\lambda_1 - \bar{\lambda}_1)f_2(\lambda_1)} = \frac{-i}{\sqrt{D_1}}\frac{f_2(\bar{\lambda}_1)(\lambda_1 - a)}{|f_2(\lambda_1)|^2} \tag{1.135}$$

$$= \frac{2}{\sqrt{D_1}}\left(\frac{(-(a_1 - a_2)a - b_1 + b_2)\sqrt{D_1}}{\left((a_1 - a_2)a_1 - 2b_1 + 2b_2\right)^2 + (a_1 - a_2)^2 D_1}\right.$$

$$\left. + i\frac{(((a_1 - a_2)a - b_1 + b_2)a_1 - 2(a_1 - a_2)b_1 + 2(b_1 - b_2)a)}{\left((a_1 - a_2)a_1 - 2b_1 + 2b_2\right)^2 + (a_1 - a_2)^2 D_1}\right),$$

since

$$f_2(\lambda_1) = -(a_1 - a_2)\lambda_1 - b_1 + b_2 \tag{1.136}$$

$$= -\frac{1}{2}\left((a_1 - a_2)a_1 + 2(b_1 - b_2) - (a_1 - a_2)\sqrt{D_1}i\right) \tag{1.137}$$

and

$$A_2 = \frac{2}{\sqrt{D_2}}\left(\frac{(-(a_2-a_1)a-b_2+b_1)\sqrt{D_2}}{((a_2-a_1)a_2-2b_2+2b_1)^2+(a_2-a_1)^2D_1}\right. \tag{1.138}$$
$$\left.+i\frac{(((a_2-a_1)a-b_2+b_1)a_2-2(a_2-a_1)b_2+2(b_2-b_1)a)}{((a_2-a_1)a_2-2b_2+2b_1)^2+(a_2-a_1)^2D_2}\right).$$

Exercise 4. Check that the Laplace inversion of (1.134) is

$$\mathcal{L}^{-1}(f)(t) = \frac{2e^{-\frac{a_1}{2}t}}{\sqrt{D_1}}\left(\frac{(-(a_1-a_2)a-b_1+b_2)\sqrt{D_1}}{((a_1-a_2)a_1-2b_1+2b_2)^2+(a_1-a_2)^2D_1}\cos\frac{\sqrt{D_1}}{2}\right.$$
$$\left.-\frac{(((a_1-a_2)a-b_1+b_2)a_1-2(a_1-a_2)b_1+2(b_1-b_2)a)}{((a_1-a_2)a_1-2b_1+2b_2)^2+(a_1-a_2)^2D_1}\sin\frac{\sqrt{D_1}}{2}\right)$$
$$\frac{2e^{-\frac{a_1}{2}t}}{\sqrt{D_2}}\left(\frac{(-(a_2-a_1)a-b_2+b_1)\sqrt{D_2}}{((a_2-a_1)a_2-2b_2+2b_1)^2+(a_2-a_1)^2D_1}\cos\frac{\sqrt{D_1}}{2}\right.$$
$$\left.-\frac{(((a_2-a_1)a-b_2+b_1)a_2-2(a_2-a_1)b_2+2(b_2-b_1)a)}{((a_2-a_1)a_2-2b_2+2b_1)^2+(a_2-a_1)^2D_2}\sin\frac{\sqrt{D_1}}{2}\right).$$

Example 1.3. Suppose $a, b, c, d \in \mathbb{R}$ satisfy $c^2 - 4d < 0$ and $b > 0$ for simplicity. Then the solution $y = y(t)$ of the DE

$$y'' + cy' + dy = e^{at}\sin bt \tag{1.139}$$

where the initial values are assumed to be 0: $y(0) = 0$, $y'(0) = 0$ is given by

$$y(t) = e^{at}\left(\frac{-(2a+c)b}{(a^2-b^2+ac+d)^2+(2a+c)^2b^2}\cos bt\right. \tag{1.140}$$
$$\left.+\frac{a^2-b^2+ac+d}{(a^2-b^2+ac+d)^2+(2a+c)^2b^2}\sin bt\right)$$
$$+\frac{4b}{\sqrt{4d-c^2}}e^{-\frac{c}{2}t}\left(\frac{(2a+c)\sqrt{4d-c^2}}{4\left(a^2+b^2+\frac{1}{2}c^2+ac\right)^2+(2a+c)^2(4d-c^2)}\cos\frac{\sqrt{4d-c^2}}{2}t\right.$$
$$\left.+\frac{2\left(a^2+b^2+\frac{1}{2}c^2+ac-d\right)}{4\left(a^2+b^2+\frac{1}{2}c^2+ac-d\right)^2+(2a+c)^2(4d-c^2)}\sin\frac{\sqrt{4d-c^2}}{2}t\right).$$

Proof. Let $Y(s) = \mathcal{L}[y](s)$ be the Laplace transform of $y(t)$. Then we have

$$\mathcal{L}[e^{at}\sin bt](s) = \frac{b}{f_1(s)} = \frac{b}{(s-\lambda_1)(s-\bar{\lambda}_1)}, \tag{1.141}$$

where $f_1(s) = s^2 - 2as + a^2 + b^2$. Also

$$f_2(s) = s^2 + cs + d = (s - \lambda_2)(s - \bar{\lambda}_2). \tag{1.142}$$

In the notation of exercises above,

$$a_1 = -2a, b_1 = a^2 + b^2, D_1 = b > 0, \lambda_1 = a + bi, \tag{1.143}$$
$$a_2 = c, b_2 = d, D_2 = 4d - c^2, \lambda_2 = \frac{-c + \sqrt{4d - c^2}}{2}$$

and we assume that $\lambda_1 \neq \lambda_2$ (the equality case is treated in). We find that

$$\frac{1}{b}Y(s) = \frac{1}{f_1(s)f_2(s)} = \frac{A_1}{s - \lambda_1} + \frac{\bar{A}_1}{s - \bar{\lambda}_1} + \frac{A_2}{s - \lambda_2} + \frac{\bar{A}_2}{s - \bar{\lambda}_2}, \tag{1.144}$$

$$A_1 = \frac{1}{2b}\frac{-(2a + c)b - i(a^2 - b^2 + ac + d)}{(a^2 - b^2 + ac + d)^2 + (2a + c)^2 b^2} \tag{1.145}$$

and

$$A_2 = \frac{2}{\sqrt{4d - c^2}}\frac{(2a + c)\sqrt{4d - c^2} - 2i\left(a^2 + b^2 + \frac{1}{2}c^2 + ac - d\right)}{4\left(a^2 + b^2 + \frac{1}{2}c^2 + ac - d\right)^2 + (2a + c)^2(4d - c^2)}. \tag{1.146}$$

Hence by the Laplace inversion

$$\frac{1}{b}y(t) = 2\,\mathrm{Re}\left(A_1 e^{\lambda_1 it}\right) + 2\,\mathrm{Re}\left(A_2 e^{\lambda_2 it}\right) \tag{1.147}$$
$$= 2e^{at}\Big(\frac{1}{2b}\frac{-(2a + c)b}{(a^2 - b^2 + ac + d)^2 + (2a + c)^2 b^2}\cos bt$$
$$+ \frac{1}{2b}\frac{a^2 - b^2 + ac + d}{(a^2 - b^2 + ac + d)^2 + (2a + c)^2 b^2}\sin bt\Big)$$
$$+ 2e^{-\frac{c}{2}t}\Bigg(\frac{2}{\sqrt{4d-c^2}}\frac{(2a + c)\sqrt{4d - c^2}}{4(-a^2+b^2-ac-d)^2+(2a+c)^2(4d-c^2)}\cos\frac{\sqrt{4d-c^2}}{2}t$$
$$+ \frac{2}{\sqrt{4d - c^2}}\frac{2\left(a^2 + b^2 + \frac{1}{2}c^2 + ac - d\right)}{4\left(a^2 + b^2\frac{1}{2}c^2 + ac - d\right)^2 + (2a + c)^2(4d - c^2)}\sin\frac{\sqrt{4d - c^2}}{2}t\Bigg)$$

or (1.140), completing the proof. □

Exercise 5. In the DE (1.123) we assume $y(x)$ is a one-dimensional wave and has the initial value $y(0) = 0$ and satisfies the boundary condition that $y(x) = 0$ for $x > a > 0$, a being big enough. Then show that (1.123) amounts to the one-dimensional Schrödinger equation

$$-\frac{\hbar}{2m^2}y'' = Ey, \tag{1.148}$$

where $E = \frac{\hbar}{2m^2}\alpha$. Find the energy levels and eigenfunctions.

Solution. By continuity, $y(a) = 0$. The solution (1.126) reads

$$y(x) = \frac{y'(0)}{\sqrt{\alpha}} \sin \sqrt{\alpha} x. \tag{1.149}$$

For this to vanish at $x = a$, we must have $\sqrt{\alpha} a = \pi n$, $n \in \mathbb{N}$ or

$$\alpha = \left(\frac{\pi n}{a}\right)^2. \tag{1.150}$$

Hence

$$E = E_n = \frac{\hbar}{2m^2}\left(\frac{\pi n}{a}\right)^2 \tag{1.151}$$

are energy levels and

$$y_n(x) = c \sin \frac{\pi n}{a} x \tag{1.152}$$

are eigenfunctions belonging to E_n.

(1.123) contains the case of **simple harmonic motion** of a point particle under the restoring force directly proportional to the displacement x acting on it in the direction opposite to x, i.e.

$$mx'' = -kx \tag{1.153}$$

which is the case (1.123) with $\alpha = \frac{k}{m}$. It is customary to write ω for the frequency, i.e. $\omega = \sqrt{\alpha} = \sqrt{\frac{k}{m}}$ or $k = m\omega^2$. Since the potential inducing the restoring force on the left-hand side of the above equation is $V(x) = \frac{1}{2}x^2$, the Hamiltonian function (1.89) is

$$H = H(x) = \frac{1}{2}mp_x^2 + \frac{1}{2}m\omega^2 x^2. \tag{1.154}$$

In case $\Psi(\boldsymbol{R}) = \Psi(x, y, z)$ in (1.98) is of variable separable type, we may separate the components [Pauling and Wilson (1935), p. 101] and it is enough to consider (1.154). Making the replacement (1.90), we find the Hamiltonian—**harmonic oscillator**

$$H = -\frac{\hbar^2}{2m}\frac{\mathrm{d}^2}{\mathrm{d}x^2} + \frac{1}{2}m\omega^2 x^2 \tag{1.155}$$

and the Schrödinger equation corresponding to (1.98) with $W = E$ reads

$$H\Psi(x) = E\Psi(x). \tag{1.156}$$

Introducing the notation

$$x \leftrightarrow \sqrt{\frac{m\omega}{\hbar}} x, \quad \lambda = \frac{2}{\hbar\omega} E,$$

(1.156) amounts to

$$\left(\frac{\mathrm{d}^2}{\mathrm{d}x^2}+x^2\right)\Psi(x)=\lambda\Psi(x) \tag{1.157}$$

which is (1.161) below satisfied by Hermite polynomials. The solutions satisfying the boundary condition that $\Psi(x)\to 0$ as $x\to\infty$ are possible only for $\lambda=\lambda_n$ with $2n+1$, $n=0,1,\cdots$ and are proportional to (1.160).

Invoking the normalization of Ψ we have proved

Theorem 1.7. *Eigenfunctions corresponding to eigenvalues* $E_n = \left(n+\frac{1}{2}\right)\hbar\omega$ *are*

$$\Psi_n(x)=\sqrt{\frac{1}{2^n n!}\sqrt{\frac{m\omega}{\pi\hbar}}}H_n\left(\sqrt{\frac{m\omega}{\hbar}}x\right)e^{-\frac{m\omega}{2\hbar}x^2}. \tag{1.158}$$

Cf. e.g. [Koide (1971)].

Here the **Hermite polynomial** $y=H_n(x)$ is a solution of the DE

$$\frac{\mathrm{d}^2y}{\mathrm{d}x^2}-2x\frac{\mathrm{d}y}{\mathrm{d}x}+2ny=0 \tag{1.159}$$

and

$$w=e^{-\frac{1}{2}x^2}H_n(x) \tag{1.160}$$

is a solution of

$$\frac{\mathrm{d}^2w}{\mathrm{d}x^2}+(2n+1-x^2)w=0. \tag{1.161}$$

Hermite polynomials are one of well-known class of orthogonal polynomials (cf. e.g. [Erdélyi *et al.* (1953), II, pp. 153-231]) and the class of functions given by (1.158) forms an ONB (orthonormal basis).

There is an explicit formula ([Erdélyi *et al.* (1953), II, p. 193 (9)])

$$H_n(x)=\sum_{m=0}^{[n/2]}\frac{(-1)^m(2x)^{n-2m}}{m!(n-2m)!} \tag{1.162}$$

which may be proved by the formula

$$H_n(x)=(-1)^n e^{x^2}D^n e^{-x^2} \tag{1.163}$$

with $D=\frac{\mathrm{d}}{\mathrm{d}x}$. First a few terms are

$$H_0(x)=1, H_1(x)=2x.H_2(x)\ 4x^2-2, H_3(x)=8x^3-12x,\cdots.$$

Exercise 6. Determine the constants α so that by the change of variable

$$y=e^{\alpha x^2}w \tag{1.164}$$

(1.159) reduces to the one without $\frac{\mathrm{d}w}{\mathrm{d}x}$ and deduce (1.161).

Solution. We have

$$\frac{\mathrm{d}y}{\mathrm{d}x} = e^{\alpha x^2}\left(2\alpha x w + \frac{\mathrm{d}w}{\mathrm{d}x}\right)$$

and

$$\frac{\mathrm{d}^2y}{\mathrm{d}x^2} = e^{\alpha x^2}\left(\frac{\mathrm{d}^2w}{\mathrm{d}x^2} + 4\alpha x\frac{\mathrm{d}w}{\mathrm{d}x} + 2\alpha\left(2\alpha x^2+1\right)w\right).$$

Substituting these and dividing by $e^{\alpha x^2}$, we see that (1.159) reads

$$\frac{\mathrm{d}^2w}{\mathrm{d}x^2} + (4\alpha - 2)\frac{\mathrm{d}w}{\mathrm{d}x} + \left(4\alpha(\alpha-1)x^2 + 2\alpha + 2n\right)w = 0. \tag{1.165}$$

We choose $\alpha = \frac{1}{2}$. Then the coefficient of w for this choice is

$$2n + 1 - x^2$$

and (1.161) follows.

The confluent hypergeometric functions are solutions of the DE

$$x\frac{\mathrm{d}^2y}{\mathrm{d}x^2} + (c-x)\frac{\mathrm{d}y}{\mathrm{d}x} - ay = 0 \tag{1.166}$$

and one of the solutions is given by $\Psi(a,c;x)$ [Erdélyi *et al.* (1953), II, p. 248].

Exercise 7. Determine the constants α, β so that by the change of variable

$$y = x^{\alpha}e^{\beta x}w \tag{1.167}$$

(1.166) reduces to the one without $\frac{\mathrm{d}w}{\mathrm{d}x}$ and deduce Whittaker's standard form

$$\frac{\mathrm{d}^2w}{\mathrm{d}x^2} + \left(-\frac{1}{4} + \frac{\varkappa}{x} + \frac{\frac{1}{4}-\mu^2}{x^2}\right)w = 0 \tag{1.168}$$

where $a = \frac{1}{2} - \varkappa + \mu$, $c = 1 + 2\mu$.

Solution. We have

$$\frac{\mathrm{d}y}{\mathrm{d}x} = x^{-\frac{c}{2}}e^{\frac{1}{2}x}\left(\left(\frac{\alpha}{x}+\beta\right)w + \frac{\mathrm{d}w}{\mathrm{d}x}\right)$$

and

$$\frac{\mathrm{d}^2y}{\mathrm{d}x^2} = x^{\alpha}e^{\beta x}\left(\frac{\mathrm{d}^2w}{\mathrm{d}x^2} + 2\left(\frac{\alpha}{x}+\beta\right)\frac{\mathrm{d}w}{\mathrm{d}x} + \left(\frac{\alpha^2-\alpha}{x^2} + \frac{2\alpha\beta}{x} + \beta^2\right)w\right).$$

Substituting these and dividing by $x^{\alpha}e^{\beta x}$, we see that (1.166) reads

$$\begin{aligned} &x\frac{\mathrm{d}^2w}{\mathrm{d}x^2} + (2(\alpha+\beta x) + c - x)\frac{\mathrm{d}w}{\mathrm{d}x} \\ &+ \left(\frac{\alpha^2-\alpha+c\alpha}{x} + 2\alpha\beta - \alpha + c\beta - a + \beta(\beta-1)x\right)w = 0. \end{aligned} \tag{1.169}$$

For the coefficient of $\frac{dw}{dx}$ to vanish, we choose $\alpha = -\frac{c}{2}$, $\beta = \frac{1}{2}$. Then the coefficient of w for this choice is

$$\frac{-\frac{c^2}{4}+\frac{c}{2}}{x}+\frac{c}{2}-a-\frac{1}{4}x.$$

Dividing (1.169) by x leads to (1.168).

Exercise 8. Show that by an appropriate change of variable, (1.166) reduces to (1.159) and (1.161).

Solution. We write the variable x by z in (1.166) and put

$$z = \alpha x^2 \tag{1.170}$$

to obtain

$$\frac{dy}{dx} = 2\alpha x\frac{dy}{dz}, \quad \frac{dy}{dz} = \frac{1}{2\alpha x}\frac{dy}{dx}.$$

Hence

$$\frac{d^2y}{dx^2} = 2\alpha\frac{dy}{dz} + 2\alpha x\frac{d}{dx}\frac{dy}{dz} = \frac{1}{x}\frac{dy}{dx} + (2\alpha x)^2\frac{d^2y}{dz^2}$$

or

$$\frac{d^2y}{dz^2} = \frac{1}{4\alpha^2x^2}\left(\frac{d^2y}{dx^2} - \frac{1}{x}\frac{dy}{dz}\right).$$

Substituting in (1.166), we deduce that

$$\frac{\alpha}{4\alpha^2}\left(\frac{d^2y}{dx^2} - \frac{1}{x}\frac{dy}{dz}\right) + (c-\alpha x^2)\frac{1}{2\alpha x}\frac{dy}{dx} - ay = 0$$

or

$$\frac{d^2y}{dx^2} + \left(c - \frac{\alpha}{2}\right)\frac{1}{2\alpha x}\frac{dy}{dx} - 2\alpha x\frac{dy}{dx} - 4\alpha ay = 0.$$

Letting $c = \frac{\alpha}{2}\alpha$, we arrive at

$$\frac{d^2y}{dx^2} - 2\alpha x\frac{dy}{dx} - 4\alpha ay = 0. \tag{1.171}$$

Hence (1.171) with $\alpha = 1$ (whence $c = \frac{1}{2}$) and $a = -\frac{1}{2}n$ amounts to (1.159).

Exercise 9. Show that ([Erdélyi *et al.* (1953), II, p. 194 (16)]

$$H_n(x) = 2^n\Psi\left(-\frac{1}{2}n, \frac{1}{2}; x^2\right). \tag{1.172}$$

Also show that [Erdélyi *et al.* (1953), I, p. 267 (32)]

$$H_n(x) = 2^n x\Psi\left(\frac{1}{2} - \frac{1}{2}n, \frac{3}{2}; x^2\right) \tag{1.173}$$

amounts to (1.172).

Solution. Recall the asymptotic formula [Erdélyi *et al.* (1953), I, p. 278 (1)]

$$\Psi(a,c;x)=\sum_{m=0}^{N}(-1)^m\frac{(a)_m(a-c+1)_m}{m!}x^{-a-m}+O\left(|x|^{-a-N-1}\right), \tag{1.174}$$

where

$$N=0,1,2,\cdots,\quad -\frac{3}{2}\pi<\arg x<\frac{3}{2}\pi,\quad |x|\to\infty.$$

Comparing the leading coefficients of (1.162) and (1.174), we conclude (1.172).

To show that (1.173) amounts to (1.172) we recall [Erdélyi *et al.* (1953), I, p. 267 (6)]

$$\Psi(a,c;x)=x^{1-c}\Psi(a-c+1,2-c:x), \tag{1.175}$$

(1.175) with $a=-\frac{n}{2}$, $c=\frac{1}{2}$, $x\leftrightarrow x^2$ establishes (1.173).

A slightly more general class of functions are that of parabolic cylinder functions, [Erdélyi *et al.* (1953), pp. 115-132] (in which the header on p. 131 should read "parabolic cylinder functions").

1.5.3 *Quantum information*

Given two wave-functions ψ_m,ψ_n, their inner product (ψ_m,ψ_n) is usually defined as in (1.78) by

$$(\psi_m,\psi_n)=\int\overline{\psi_m}\psi_n\,\mathrm{d}\tau. \tag{1.176}$$

We show in (1.184) below that

$$(\psi_m,\psi_n)=\langle\psi_m|\psi_n\rangle \tag{1.177}$$

in the Dirac notation [Dirac (1958)].

Let $\{u_i\}$ be an ONB (e.g. (1.158) in infinite dimensional case). Then an arbitrary wave-function may be expressed as

$$\psi_m=\sum_i c_{im}u_i, \tag{1.178}$$

where $c_{im}=(\psi_m,u_i)\in\mathbb{C}$ such that

$$\sum_i|c_{im}|^2=1. \tag{1.179}$$

We define the **ket vector** $|\psi_m\rangle$ by

$$|\psi_m\rangle=\begin{pmatrix}c_{1m}\\c_{2m}\\\vdots\end{pmatrix} \tag{1.180}$$

and the **bra vector** $\langle\psi_m|$ by

$$\langle\psi_m| = |\psi_m\rangle^* = \left(\overline{c_{1m}}\ \overline{c_{2m}}\ \cdots\right), \tag{1.181}$$

where $*$ means the *conjugation* (defined by (1.77)) and $^-$ the complex conjugation, cf. Table 1.2 below.

(1.180) reads for the element u_i of the ONB,

$$|u_i\rangle = \begin{pmatrix} 0 \\ 0 \\ \vdots \\ 1 \\ 0 \\ \vdots \end{pmatrix} = \boldsymbol{e}_i, \tag{1.182}$$

the ith fundamental unit vector. (1.179) implies that

$$\Big\| |\psi_m\rangle \Big\| = 1. \tag{1.183}$$

Substituting (1.179) into (1.176), we find that

$$\begin{aligned}(\psi_m, \psi_n) &= \sum_i \sum_j \overline{c_{im}} c_{jn} (u_i u_j) = \sum_i \sum_j \overline{c_{im}} c_{jn} \delta_{ij} \\ &= \sum_i \overline{c_{im}} c_{in} = (\langle\psi_m|, |\psi_n\rangle),\end{aligned} \tag{1.184}$$

i.e. the inner product of the bra and ket-vectors, which we express as $\langle\psi_m|\psi_n\rangle$ following Dirac's convention [Dirac (1958), p. 19, ll. 15-16 from below] of contracting two vertical lines into one. Hence (1.184) leads to (1.177). Up here the Hilbert spaces comprising of wave functions are of infinite dimensional and are referred to as **continuous-variable quantum information**. The ket vectors are of infinite degree, which expression is rather classical, cf. e.g. [Wintner (1925)] and infinite degree matrices are not often used nowadays. For modern theory cf. e.g. [Braunstein and Pati (2012)].

In what follows we shall confine ourselves to qubits where the associated Hilbert spaces are finite dimensional complex vector spaces $\mathcal{H} \simeq \mathbb{C}^N$ (cf. (1.191)). Hence a ket vector is an N-dimensional column vector and ordinary linear algebra applies with *all operators being matrices*. Therefore $*$ means the *transposed bar operation* and $^-$ the complex conjugation, i.e.

names	complex conj.	transpose	transp. conj.
math.	$\bar{A}$	${}^tA$	A^*
quantum th.	A^*	${}^TA$	$A^\dagger$

Table 1.2: Symbols

$\bar{A} = (\bar{a}_{ij})$ and $A^* = {}^t\bar{A}$. According to §1.5, we are mainly concerned with Hermite matrices (Hamiltonians) $A^* = A$ and unitary matrices $U^* = U^{-1}$. As is defined below, qubits are (N-dimensional) unit vectors, unitary matrices are essential because they are *isometries*: $\big|U|\psi\rangle\big| = \big||\psi\rangle\big|$.

From here to the end of this section depends partly on [Nakahara and Ohimi (2008)], [Nielsen and Chuang (2010)] and some materials are taken from there with more detailed and more lucid expoundation using more well-known symbols from linear algebra. It is standard to start from the 2-dimensional complex vector space $\mathcal{H} \simeq \mathbb{C}^2 = \mathbb{C}\boldsymbol{e}_1 \oplus \mathbb{C}\boldsymbol{e}_2$. It is customary to denote the unit basis vectors as two special states $|0\rangle, |1\rangle$ known as *computational basis states*. In the notation of (1.182)

$$|0\rangle = \begin{pmatrix}1\\0\end{pmatrix} = \boldsymbol{e}_1, \quad |1\rangle = \begin{pmatrix}0\\1\end{pmatrix} = \boldsymbol{e}_2. \tag{1.185}$$

These correspond to the spin-up and spin-down states of a particle such as an electron or a proton. All the elements $a|0\rangle + b|1\rangle$, $|a|^2 + |b|^2 = 1$ in $\mathcal{H} \simeq \mathbb{C}^2$ are called (single) **qubits** (quantum bits) or a 1-qubit. When qubits are *observed* (or measured), the value is determined either to be 0 or 1 with probability $|a|^2$ and $|b|^2$, respectively. Measurement is formally conducted by applying the measurement operators in Definition 1.5 and the coefficients $|a|^2$ and $|b|^2$ may be thought of as probabilities (cf. Remark 1.4).

Single qubits are extended to ν-qubits by way of tensor products.

Definition 1.2. For two free $\mathbb{C}$-modules M, N (i.e. $\mathbb{C}$-vector spaces) with bases $\{u_\mu\}$, $\{v_\nu\}$, their tensor product $M \otimes N$ is the totality of all linear combinations of $u_\mu \otimes v_\nu$, which is again a free $\mathbb{C}$-module. For matrices, we specify the tensor product to be the Kronecker product in Definition 1.3. Hence 2-qubits are all the unit vectors in the state space $\mathbb{C}^2 \otimes \mathbb{C}^2$. Then we define ν-qubits inductively as unit vectors in (1.188) below.

For linear maps $f : M \to M'$, $g : N \to N'$, there exists a unique linear map denoted $f \otimes g$ such that

$$(f \otimes g)(x \otimes y) = f(x) \otimes g(y) \tag{1.186}$$

and it holds that

$$(f' \otimes g') \circ (f \otimes g) = (f' \circ f) \otimes (g' \circ g), \quad I_M \otimes I_N = I_{M \otimes N}, \tag{1.187}$$

where I_M indicates the identity map.

ν-qubits are all the unit vectors in the Hilbert space (state space)

$$\mathcal{H} \simeq \underbrace{\mathbb{C}^2 \otimes \cdots \otimes \mathbb{C}^2}_{\nu}, \tag{1.188}$$

i.e. they are of the form

$$a_1 \boldsymbol{e}_1 + \cdots + a_\nu \boldsymbol{e}_N \tag{1.189}$$

with normalization

$$\sum_{k=1}^{N} |a_k|^2 = 1, \tag{1.190}$$

where

$$\boldsymbol{e}_1 = |0 \cdots 0\rangle, \cdots, \boldsymbol{e}_N = |1 \cdots 1\rangle$$

and $|0 \cdots 0\rangle = \underbrace{|0\rangle \otimes \cdots \otimes |0\rangle}_{\nu}$ is the Kronecker product of ν $|0\rangle$'s. We write

$$N = 2^\nu \tag{1.191}$$

unless otherwise stated.

Definition 1.3. For any matrices $A = (a_{ij})_{1 \le i \le m, 1 \le j \le n}$ and $B = (a_{k\ell})_{1 \le k \le p, 1 \le \ell \le q}$ their **Kronecker product** (or tensor product) $A \otimes B$ is defined by

$$A \otimes B = \begin{pmatrix} a_{11}B & \cdots & a_{1n}B \\ & \cdots & \\ a_{m1}B & \cdots & a_{mn}B \end{pmatrix} = (a_{ij} b_{k\ell}) \tag{1.192}$$

and this extends linearly in both variables. In particular for two ket vectors

$$|\varphi\rangle = \begin{pmatrix} c_1 \\ c_2 \\ \vdots \\ c_m \end{pmatrix} \in \mathbb{C}^m, \quad |\psi\rangle = \begin{pmatrix} d_1 \\ d_2 \\ \vdots \\ d_n \end{pmatrix} \in \mathbb{C}^n \tag{1.193}$$

their Kronecker product is

$$|\varphi\rangle \otimes |\psi\rangle = (c_i d_j) = \begin{pmatrix} c_1 d_1 \\ \vdots \\ c_1 d_n \\ \\ \vdots \\ c_m d_1 \\ \vdots \\ c_m d_n \end{pmatrix}, \tag{1.194}$$

which we usually express in the transposed row vector form

$$^t(c_1 d_1, \cdots, c_1 d_n, \cdots, c_m d_1, \cdots, c_m d_n).$$

The left-hand side is abbreviated as $|\varphi\rangle|\psi\rangle = |\varphi\psi\rangle$ which satisfies

$$\big||\varphi\rangle \otimes |\psi\rangle\big| = \big||\varphi\rangle\big|\big||\psi\rangle\big|. \tag{1.195}$$

The Kronecker product $|\varphi\rangle^* \otimes |\psi\rangle$ is defined to be the **outer product** $|\psi\rangle\langle\phi|$ given by

$$|\psi\rangle\langle\varphi| = |\varphi\rangle^* \otimes |\psi\rangle = (\bar{c_1}, \cdots, \bar{c_m}) \otimes \begin{pmatrix} d_1 \\ d_2 \\ \vdots \\ d_n \end{pmatrix} = \begin{pmatrix} \bar{c_1} d_1 & \cdots & \bar{c_m} d_1 \\ & \cdots & \\ \bar{c_1} d_n & \cdots & \bar{c_m} d_n \end{pmatrix}. \tag{1.196}$$

It is convenient to view it as a linear map $\mathbb{C}^m \to \mathbb{C}^n$

$$|\psi\rangle\langle\varphi|(|\varphi'\rangle) = |\psi\rangle\langle\varphi|\varphi'\rangle = \langle\varphi|\varphi'\rangle|\psi\rangle. \tag{1.197}$$

If $\{e_j\}$ be an ONB, then the outer product

$$P_k = |e_k\rangle\langle e_k| \tag{1.198}$$

is called a **projector**, cf. [Li *et al.* (2017), p. 173].

Example 1.4. 3-qubits can represent 2^3 states. E.g. by (1.194)

$$|5\rangle = |101\rangle = |1\rangle|0\rangle|1\rangle = |1\rangle \otimes |0\rangle \otimes |1\rangle \tag{1.199}$$

$$= \begin{pmatrix} 0 \\ 1 \end{pmatrix} \otimes \begin{pmatrix} 1 \\ 0 \end{pmatrix} \otimes \begin{pmatrix} 0 \\ 1 \end{pmatrix} = \begin{pmatrix} 0 \\ 0 \\ 1 \\ 0 \end{pmatrix} \otimes \begin{pmatrix} 0 \\ 1 \end{pmatrix} = {}^t(0,0,0,0,0,1,0,0) = \boldsymbol{e}_6.$$

Here the first notation is the decimal number 5 viewed as its dyadic expansion $2^2 + 0 \cdot 2 + 1$. In general, we have

$$|n\rangle = \boldsymbol{e}_{n+1}, \tag{1.200}$$

where the left-hand side means the mth member of the sequence of N digits numbers in dyadic expansion arranged in increasing order and the right means the mth N-dimensional fundamental unit vector.

Exercise 10. Check that (1.196) and (1.197) are equivalent. Let $\{e_j\}$ be an ONB satisfying $\langle e_i|e_j\rangle = \delta_{ij}$. Then show that the completeness condition

$$\sum_{i=1}^{n} |e_i\rangle\langle e_i| = E \tag{1.201}$$

as well as

$$P_k^2 = P_k, \quad \langle P_k\varphi_1|\varphi_2\rangle = \langle\varphi_1|P_k\varphi_2\rangle, \quad P_kP_\ell = 0 \quad (k \neq \ell) \tag{1.202}$$

are satisfied, where P_k is the projector defined by (1.198).

Solution. Let $|\varphi'\rangle = \begin{pmatrix} c_1' \\ c_2' \\ \vdots \\ c_m' \end{pmatrix}$. Then noting $\langle\varphi|\varphi'\rangle = \sum_{i=1}^{m} \bar{c}_i c_i'$, we see the equivalence by multiplying $|\varphi'\rangle$ by the matrix (1.196).

To prove the second assertion, we note a general formula

$$|e_i\rangle\langle e_j| = E_{ij}, \tag{1.203}$$

where E_{ij} is a matrix unit consisting of all 0's except for the (i, j)-entry which is 1. Finally the second equality in (1.202) holds since both sides are $\langle e_k|\varphi_1\rangle\langle e_k|\varphi_2\rangle$.

Exercise 11. Prove that the totality of all Kronecker products of single qubits $|a\rangle|b\rangle$ does not exhaust the totality of 2-qubits states. As an example $\frac{|00\rangle+|11\rangle}{\sqrt{2}}$.

Solution. 2-qubits are of the form

$$a_1|00\rangle + a_2|01\rangle + a_3|10\rangle + a_4|11\rangle \tag{1.204}$$

with $\sum_{k=1}^{4} |a_k|^2 = 1$.

On the other hand, by (1.200), we have, with the second term indicating the dyadic expansion,

$$\begin{aligned} |0\rangle \otimes |0\rangle &= |0\rangle = \boldsymbol{e}_1, \quad & |0\rangle \otimes |1\rangle &= |1\rangle = \boldsymbol{e}_2, \\ |1\rangle \otimes |0\rangle &= |2\rangle = \boldsymbol{e}_3, \quad & |1\rangle \otimes |1\rangle &= |3\rangle = \boldsymbol{e}_4. \end{aligned} \tag{1.205}$$

Hence

$$|a\rangle \otimes |b\rangle = (a_1|0\rangle + b_1|1\rangle) \otimes (a_2|0\rangle + b_2|1\rangle) = {}^t(a_1a_2, a_1b_2, a_2b_1, b_1b_2), \quad (1.206)$$

which cannot exhaust (1.204).

E.g.

$$\frac{|00\rangle + |11\rangle}{\sqrt{2}} = {}^t\left(\frac{1}{\sqrt{2}}, 0, 0, \frac{1}{\sqrt{2}}\right) \quad (1.207)$$

cannot be expressed in the form (1.206).

As is stated in Definition 1.2 above, the state space of a composite quantum system is the tensor product of the state spaces of the component systems. If $|\psi_j\rangle$, $1 \le j \le n$ are the states of the component systems (N_j-qubit), then the joint state of the total system is the Kronecker product $|\psi_1\rangle \otimes \cdots \otimes |\psi_n\rangle$, which is an $N = N_1 \cdots N_n$-qubit in $\mathbb{C}^N$. For the length of this vector is $\big||\psi_1\rangle\big| \cdots \big||\psi_n\rangle\big| = 1$ in view of (1.195). But since the totality of all N-qubits arising from the Kronecker products of component N_j-qubits have restrictions on the coefficients, it does not exhaust all N-qubits. In qubit theory one says that a state of a composite system which cannot be written as a product of states of its component systems is an **entangled state**. The state in Exercise 1.4 is an entangled state.

It is known that a necessary and sufficient condition for a square matrix A to be diagonalizable by a unitary matrix is that A is a normal matrix $A^*A = AA^*$. Hence an Hermite matrix is diagonalizable by a unitary matrix. It follows that if the eigenvalues of A are λ_k, $1 \le k \le n$ we may always choose the set of eigenvectors forming an ONB of $\mathbb{C}^n$.

Theorem 1.8. *Let A be an Hermite matrix of degree n with eigenvalues λ_k and the corresponding eigenvectors $|\lambda_k\rangle$, $1 \le k \le n$ forming an ONB. Then the spectral decomposition*

$$A = \sum_{k=1}^{n} \lambda_k P_k = \sum_{k=1}^{n} \lambda_k |\lambda_k\rangle\langle\lambda_k| \quad (1.208)$$

holds true. If $f(z) = \sum_{k=0}^{\infty} a_k z^k$ is an analytic function with power series convergent in a disc, then

$$f(A) = \sum_{k=1}^{n} f(\lambda_k) |\lambda_k\rangle\langle\lambda_k| \quad (1.209)$$

and in particular,

$$e^A = \sum_{k=1}^{n} e^{\lambda_k} |\lambda_k\rangle\langle\lambda_k|. \quad (1.210)$$

For any $|a\rangle = \sum_{\ell=1}^{n} a_\ell|\lambda_\ell\rangle$ we have

$$\sum_{k=1}^{n} \lambda_k P_k|a\rangle = \sum_{k=1}^{n} \lambda_k|\lambda_k\rangle\langle\lambda_k| \sum_{\ell=1}^{n} a_\ell|\lambda_\ell\rangle = \sum_{k,\ell} a_\ell\lambda_k|\lambda_k\rangle\langle\lambda_k|\lambda_\ell\rangle = \sum_{k=1}^{n} a_k\lambda_k|\lambda_k\rangle \tag{1.211}$$

in view of orthogonality. Since $A|a\rangle = \sum_{k=1}^{n} a_k A|\lambda_k\rangle = \sum_{k=1}^{n} a_k\lambda_k|\lambda_k\rangle$, (1.208) follows. Cf. [Li *et al.* (2017), p. 184] for a spectral decomposition for a self-adjoint operator.

Theorem 1.8 is a special case of Lemma 1.3 in the case of a Hermite matrix. Indeed, writing

$$A = UDU^* = \sum_{k=1}^{n} \lambda_k U E_{kk} U^*, \tag{1.212}$$

we see that this amounts to (1.208).

Exercise 12. Let

$$\sigma_y = \sigma_2 = \begin{pmatrix} 0 & -i \\ i & 0 \end{pmatrix} \tag{1.213}$$

be one of three three Pauli matrices. Then prove that

$$e^{i\gamma\sigma_y} = \begin{pmatrix} \cos\gamma & -\sin\gamma \\ \sin\gamma & \cos\gamma \end{pmatrix} \tag{1.214}$$

which is the matrix of rotation in Theorem 1.3.

Solution. It is easy to see that the eigenvalues and the corresponding eigenspaces of σ_y are $\lambda_1 = 1, \lambda_2 = -1$, $E_{\sigma_y}(1) = \mathbb{C}\begin{pmatrix} -i \\ 1 \end{pmatrix}$, $E_{\sigma_y}(-1) = \mathbb{C}\begin{pmatrix} i \\ 1 \end{pmatrix}$. With $|\lambda_1\rangle = \frac{1}{\sqrt{2}}\begin{pmatrix} -i \\ 1 \end{pmatrix}$, $|\lambda_2\rangle = \frac{1}{\sqrt{2}}\begin{pmatrix} i \\ 1 \end{pmatrix}$, we have

$$|\lambda_1\rangle\langle\lambda_1| = \frac{1}{2}\begin{pmatrix} 1 & -i \\ i & 1 \end{pmatrix}, \quad |\lambda_2\rangle\langle\lambda_2| = \frac{1}{2}\begin{pmatrix} 1 & i \\ -i & 1 \end{pmatrix}. \tag{1.215}$$

Substituting this in (1.210), we conclude (1.214). This can also be proved by using Lemma 1.3 with $P = \frac{1}{2}\begin{pmatrix} -i & i \\ i & 1 \end{pmatrix}$.

Information processing (IP) f is a process by which an input data is processed by f and an output data is released. This has been done by classical computers whose circuits consist of wires and logical gates, where the former carry information around the circuit and the latter perform manipulations of information from one form to another. A parallel information processing (PIP) is the case where there are several processors f_i working on an input to give an output, where processors are not necessarily independent. Quantum information processing (QIP) is similar in spirit to PIP in that instead of several processors one uses only one processor—a (programmed) unitary operator as in (1.92) and the logical gate performing this *operation of multiplying by a unitary matrix* is called a **quantum gate**:

$$|\Psi_{\rm out}\rangle = U_f|\Psi_{\rm in}\rangle, \quad U_f = e^{-\frac{i}{\hbar}H_f t}, \tag{1.216}$$

where H_f is the Hamiltonian corresponding to the processor f. The quantum computer to be defined in Definition 1.6 drastically reduces the number of computational steps by means of the generated entanglement.

It is noticeable that quantum physics, esp. quantum entanglement has a close connection with the Riemann zeta-function, esp. the Riemann-Siegel formula, [Feiler and Schreich (2013)], [Feiler (2013)], [Latorre and Sierra (2015)], [Mack *et al.* (2010)], [Torosov *et al.* (2013)] etc.

Example 1.5. The NOT gate acting on single qubits is the one which changes $|0\rangle$ and $|1\rangle$ and can be represented by the matrix

$$X = \begin{pmatrix} 0 & 1 \\ 1 & 0 \end{pmatrix} \tag{1.217}$$

which is a unitary matrix and one of Pauli matrices denoted $\sigma_1 = \sigma_x = X$.

Let U be a unitary matrix of degree 2. Then the **controlled U-gate** acting on 2-qubits is the one which transforms the second bit (target bit) by U only when the first bit (the control bit) is $|1\rangle$.

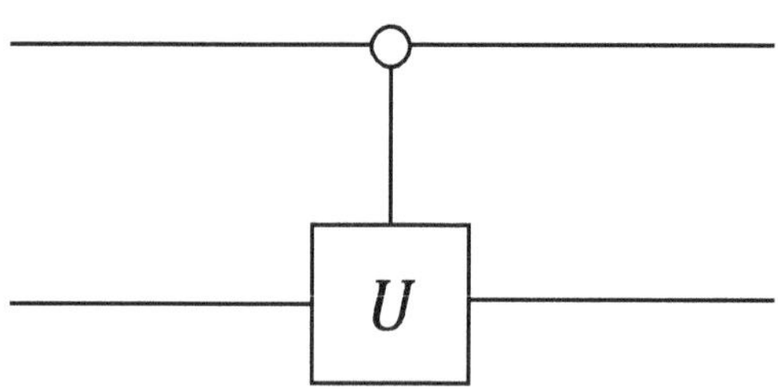

Fig. 1.2: Controlled U-gate

In the special case where the unitary matrix U is given by

$$U_{\mathrm{CNOT}} = \begin{pmatrix} 1 & 0 & 0 & 0 \\ 0 & 1 & 0 & 0 \\ 0 & 0 & 0 & 1 \\ 0 & 0 & 1 & 0 \end{pmatrix} = \begin{pmatrix} E_2 & O \\ O & X \end{pmatrix}, \tag{1.218}$$

the controlled U-gate is called the **CNOT** (controlled NOT gate). I.e. CNOT gate acts on 2-qubits and changes $|xy\rangle$ into $|x(x+y) \mod 2\rangle$, i.e. it changes the target bit—the second bit by the other one only when the control bit—the first bit is 1.

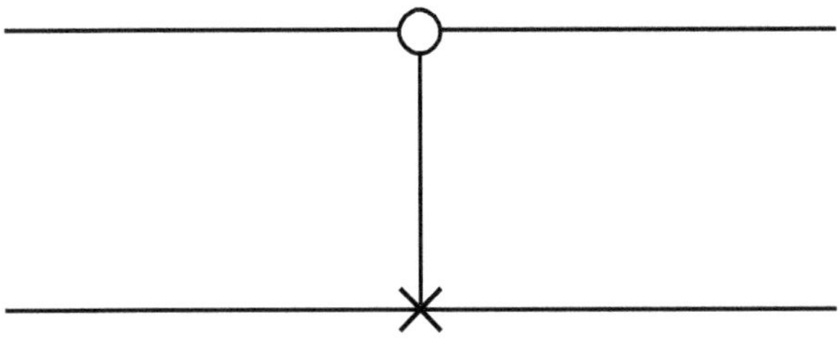

Fig. 1.3: CNOT gate

For any unitary matrices used U_1, U_2, the conditional transformation

$$U = |0\rangle\langle 0| \otimes U_1 + |1\rangle\langle 1| \otimes U_2 \tag{1.219}$$

is also unitary. For

$$UU^* = (|0\rangle\langle 0| \otimes U_1 + |1\rangle\langle 1| \otimes U_2)(|0\rangle\langle 0| \otimes U_1^* + |1\rangle\langle 1| \otimes U_2^*)$$

reduces to $|0\rangle\langle 0| \otimes E_2 + |1\rangle\langle 1| \otimes E_2 = E_2 \otimes E_2 = E_4$ since

$$E_{ii}E_{jj} = \delta_{ij}E_{ij}. \tag{1.220}$$

Since

$$\begin{pmatrix} E_2 & O \\ O & X \end{pmatrix} = \begin{pmatrix} E_2 & O \\ O & O \end{pmatrix} + \begin{pmatrix} O & O \\ O & X \end{pmatrix} = E_{11} \otimes E_2 + E_{22} \otimes X = |0\rangle\langle 0| \otimes E_2 + |1\rangle\langle 1| \otimes X, \tag{1.221}$$

it follows that the matrix U_{CNOT} in (1.218) is a conditional transformation and *a fortiori* is a unitary matrix.

Example 1.6. The Hadamard gate H is defined by

$$H|x\rangle = \frac{1}{\sqrt{2}} \sum_{y=0}^{1} (-1) - xy|y\rangle. \tag{1.222}$$

Hence it sends $|0\rangle$ (resp. $|1\rangle$) to $\frac{1}{\sqrt{2}}(|0\rangle + |1\rangle)$ (resp. $\frac{1}{\sqrt{2}}(|0\rangle - |1\rangle)$) and is represented by the matrix $H = \frac{1}{\sqrt{2}}\begin{pmatrix} 1 & 1 \\ 1 & -1 \end{pmatrix}$. Since $H^2 = E_2$ we have

$$HH|0\rangle = |0\rangle, \quad HH|1\rangle = |1\rangle. \tag{1.223}$$

Consider the CNOT gate with the control and target bits are squeezed by two Hadamard gates. Then if the first input bit is $|1\rangle$, then by the conditional negation, the second input bit is interchanged, i.e. the second input bit after the controlled negation becomes $H|1\rangle$ (resp. $H|0\rangle$) if the second input bit is $H|0\rangle$ (resp. $H|1\rangle$). Then this is processed by the last Hadamard gate to become $H|1\rangle$ (resp. $H|0\rangle$), i.e. this circuit has the same effect as the up-side-down CNOT gate.

We have also

$$H \otimes H|00\rangle = H \otimes H|0\rangle \otimes 0\rangle = H|0\rangle \otimes H|0\rangle = \frac{1}{\sqrt{2}}(|0\rangle + |1\rangle) \otimes \frac{1}{\sqrt{2}}(|0\rangle + |1\rangle). \tag{1.224}$$

Let

$$U = \begin{pmatrix} E_2 & O \\ O & \tilde{U} \end{pmatrix} \tag{1.225}$$

where

$$\tilde{U} = \begin{pmatrix} a & \boldsymbol{o} & b \\ \boldsymbol{o} & E_4 & \boldsymbol{o} \\ c & \boldsymbol{o} & d \end{pmatrix} \tag{1.226}$$

where $a, b, c, d \in \mathbb{C}$ satisfy the conditions under which $\tilde{U}$ is unitary. Let

$$\tilde{\tilde{U}} = \begin{pmatrix} a & b \\ c & d \end{pmatrix}. \tag{1.227}$$

U acts non-trivially only on the 3rd qubit $\alpha|010\rangle$ and the 8th qubit $\beta|111\rangle$ of a 3-qubit giving rise to

$$^t(0, 0, a\alpha + b\beta, 0, 0, 0, 0, c\alpha + d\beta).$$

We interpret this as an action of $\tilde{U}$. A Grey code connecting $|010\rangle$ and $|111\rangle$ is

$$g_1 = |010\rangle \to g_2 = |011\rangle \to g_3 = |111\rangle. \tag{1.228}$$

Since g_2 and g_3 are different only at the first qubit, we think of them as contracted to $\alpha|0\rangle + \beta|1\rangle = \begin{pmatrix} \alpha \\ \beta \end{pmatrix}$ and multiply it by $\tilde{\tilde{U}}$ and then pull back the first entry $a\alpha + b\beta$ to the 3rd qubit $|010\rangle$-position as shown in Fig. 1.4.

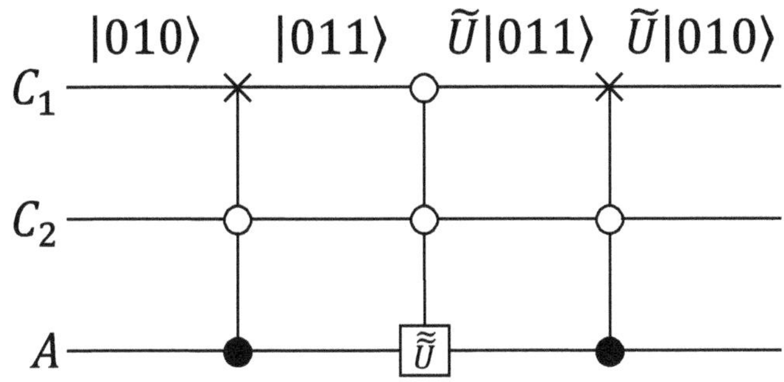

Fig. 1.4: Circuit implementing U in (1.225)

It can be shown that any unitary matrix U can be expressed as the product of level 2-matrices. This is a step toward the proof of Theorem 1.9 at the end. For its proof we refer to [Nakahara and Ohimi (2008)].

The theory of finite Fourier transforms (or discrete Fourier transform) has been developed in [Li *et al.* (2012)] in the case of periodic functions and in [Li *et al.* (2017), pp. 109-114] in the case of a finite group. Cf. also [Kanemitsu and Tsukada (2007), §8.1], [Chakraborty *et al.* (2009), §4.1, 4.3].

Definition 1.4. Let

$$\varepsilon_j(x) = e^{2\pi i x/N}, \quad 1 \le j \le N, \tag{1.229}$$

where x is an integer variable and N is defined by (1.191). Then the set $\{\varepsilon_j(x)|1 \le j \le N\}$ forms a basis of the vector space $C(N)$ of all periodic arithmetic functions with period N, where an arithmetic function is one defined only for integer arguments. We define the discrete Fourier transform (DFT) $\hat{f}$ (or the yth Fourier coefficient) of f by

$$\hat{f}(y) = \frac{1}{\sqrt{N}} \sum_{x=0}^{N-1} \varepsilon_y(-x) f(x). \tag{1.230}$$

Then the Fourier inversion or Fourier expansion formula holds true:

$$f(x) = \sum_{y=0}^{N-1} \hat{f}(y) \varepsilon_y(-x). \tag{1.231}$$

Note that (1.231) is the expression of f with respect to the basis $\{\varepsilon_j\}$.

Example 1.7. The case $N = 2$ of (1.230) amounts to the Hadamard gate. This is so since $\varepsilon_j(x) = (-1)^{xj}$ and the quantum gate realizing the DFT ($N = 2$) coincides with (1.222).

Definition 1.5. Let

$$u_1, \cdots, u_N \qquad (N = 2^\nu) \tag{1.232}$$

be an ONB of $\mathcal{H}$ and let

$$|\psi\rangle = \sum_{m=1}^{N} a_m |u_m\rangle, \tag{1.233}$$

where $a_m = \langle\psi|u_m\rangle$. Then we define the **measurement operator** M_m by

$$M_m|\psi\rangle = a_m|u_m\rangle. \tag{1.234}$$

After the measurement by M_m, the probability that the result involves $|u_m\rangle$ is

$$p(m) = \langle\psi|M_m^* M_m|\psi\rangle = (M_m|\psi\rangle, M_m|\psi\rangle) = |M_m|\psi\rangle|^2. \tag{1.235}$$

Normalizing the vector in (1.234), we see that the state after the measurement is to be

$$\frac{M_m|\psi\rangle}{\sqrt{p(m)}} = \frac{a_m}{\sqrt{p(m)}}|u_m\rangle \tag{1.236}$$

whose right-hand side is sometimes abbreviated as $|m\rangle$ to indicate that the state after the measurement is the mth basis element, so that more information is contained than (1.200).

Remark 1.4. Since

$$\langle\psi|M_m^* = |\psi\rangle^* M_m^* = (M_m|\psi\rangle)^* = (a_m|u_m\rangle)^* = \bar{a}_m|u_m\rangle \tag{1.237}$$

by (1.234), it follows that

$$p(m) = |a_m|^2\langle u_m|u_m\rangle = |a_m|^2. \tag{1.238}$$

This explains the statement after (1.185) about the probability of a qubit $a|0\rangle + b|1\rangle$.

(1.84) is viewed as the probability, which is the inner product $\langle\psi|\psi\rangle$. Since (1.235) is also an inner product $\langle M_m\psi|M_m\psi\rangle$ by (1.235), it is reasonable to view (1.235) as the probability.

Example 1.8. (i) By (1.203),

$$M_1 = M_{|0\rangle} = |0\rangle\langle 0|, \quad M_2 = M_{|1\rangle} = |1\rangle\langle 1| \tag{1.239}$$

work as measurement operators on 1-qubits: If $|\psi\rangle = a|0\rangle + b|1\rangle$ then $p(0) = |a|^2, p(1) = |b|^2$, so that the state after the measurement (1.236) reads

$$\frac{M_{|0\rangle}|\psi\rangle}{|a|} = \frac{a}{|a|}|0\rangle, \quad \frac{M_{|1\rangle}|\psi\rangle}{|b|} = \frac{b}{|b|}|1\rangle. \tag{1.240}$$

(ii) Similarly, the measurement operators acting on the first qubit of a 2-qubit given by (1.204) are

$$M_1 = |0\rangle\langle 0| \otimes E_2 = \begin{pmatrix} E_{11} & O \\ O & E_{11} \end{pmatrix}, \quad M_2 = |1\rangle\langle 1| \otimes E_2 = \begin{pmatrix} E_{22} & O \\ O & E_{22} \end{pmatrix}. \tag{1.241}$$

Expressing (1.204) as

$$|\psi\rangle = |0\rangle \otimes (a_1|0\rangle + a_2|1\rangle) + |1\rangle \otimes (a_3|0\rangle + a_4|1\rangle) \tag{1.242}$$

we obtain

$$\begin{aligned} M_1|\psi\rangle &= (|0\rangle\langle 0| \otimes E_2)\,(|0\rangle \otimes (a_1|0\rangle + a_2|1\rangle)) \\ &= |0\rangle\langle 0|\,(|0\rangle) \otimes E_2\,(a_1|0\rangle + a_2|1\rangle) \\ &= |0\rangle \otimes (a_1|0\rangle + a_2|1\rangle) \end{aligned} \tag{1.243}$$

with $p(0) = |a_1|^2 + |a_2|^2$. Hence the state after the measurement (1.236) reads

$$\frac{M_1|\psi\rangle}{\sqrt{p(0)}} = |0\rangle \otimes \left(\frac{a}{\sqrt{p(0)}}|0\rangle + \frac{b}{\sqrt{p(0)}}|1\rangle \right) \tag{1.244}$$

and similarly for $\frac{M_1|\psi\rangle}{\sqrt{p(1)}}$. The measurement operators acting on the second qubit can be introduced verbatim.

Definition 1.6. Let $\mathcal{H}$ be the state space of a quantum system, i.e. the Hilbert space in (1.188), U be a unitary operator acting on $\mathcal{H}$ and let $\{M_m\}$ be a set of measurement operators. The triple $(\mathcal{H}, U, \{M_m\})$ is called a **quantum computer**. In this context, $\mathcal{H}$ is called a register and U an algorithm.

Theorem 1.9. (Barenco *et al.*) *The set of single qubit gates and the* CNOT *gate is universal, i.e. any quantum circuit or any unitary matrix can be constructed by them.*

There are a few well-known quantum algorithms which enable to accomplish computation not possible by conventional computers in polynomial time. For well-known Shor or Glover algorithms, cf. e.g. [Nakahara and Ohimi (2008)].

1.6 Appendix A: Differential equations

Any equation in an unknown function y is called a **differential equation** abbreviated as DE if it contains derivatives of y. Since many natural phenomena obey some regular rules locally (in an infinitesimal period of time), DE is essential in studying such phenomena.

The following proposition and theorem depending on $D^{-1}D = I$ give a basis for all exponential phenomena.

Proposition 1.1. *The solution of the homogeneous DE (***autonomous equation***) in $x = x(t)$ for the* **first order reaction**

$$\frac{\mathrm{d}}{\mathrm{d}t}x = ax, \tag{1.245}$$

where a is a constant, is given by

$$x(t) = e^{at}x(0). \tag{1.246}$$

Proof. The first proof uses the fact that (1.245) is of **variable separable type**. Supposing $x(t) \neq 0$, we obtain $\frac{x'}{x} = adt$. Integrating this by means of the inverse of the logarithmic differentiation, we obtain $\log|x| = at + C_1$. Exponentiating this, we have

$$x = Ce^{at}. \tag{1.247}$$

Putting $t = 0$, we see that $C = x(0)$. Hence (1.246) follows. The case $x = 0$ corresponds to $C = 0$.

The second proof turns out to be very useful not only for this but also for other types of DEs. We must predict the final form of the solution and dividing $x(t)$ by e^{at}. The derivative of xe^{-at} is $e^{-at}(x' - ax)$ whose second factor is 0 because of (1.245). Hence in view of the Newton-Leibniz rule, we conclude that $xe^{-at} = C$, which leads to (1.247).

The third proof uses the Laplace transform. □

We illustrate the method used in the second proof above of differentiating the product of the solution and the inverse of the predicted exponential function (and using the Newton-Leibniz rule or integration) by the following Proposition 1.10, which is a generalization of Proposition 1.1:

Theorem 1.10. *The solution of the inhomogeneous linear DE*

$$\frac{\mathrm{d}}{\mathrm{d}t}x - ax = u, \tag{1.248}$$

in the **state function** $x = x(t)$, *where* $u = u(t)$ *is a given function (***input function***) and* a, b *are constants, is given by*

$$x = x(t) = e^{at}x(0) + e^{at}\int_0^t e^{-a\tau}u(\tau)\,\mathrm{d}\tau. \tag{1.249}$$

Proof. We differentiate xe^{-at} to obtain $e^{-at}(x' - ax)$, which is $e^{-at}u$ in view of (1.248):

$$(xe^{-at})' = e^{-at}u. \tag{1.250}$$

Integration of (1.250) gives

$$e^{-at}x = C + \int_0^t e^{-a\tau}u(\tau)\,\mathrm{d}\tau. \tag{1.251}$$

The integration constant C is determined to be $x(0)$, so that (1.251) leads to (1.249), completing the proof. □

We give a solution to the system of DEs with constant coefficients by linear algebra, the Laplace transform method being given in §1.5.2. The first-order reaction will appear again in §5.1.

First we consider the inhomogeneous system of DEs $\frac{\mathrm{d}}{\mathrm{d}t}\boldsymbol{x} - A\boldsymbol{x} = \boldsymbol{b}$:

$$\begin{cases} \frac{\mathrm{d}x_1}{\mathrm{d}t} = ax_1 + bx_2 + b_1 \\ \frac{\mathrm{d}x_2}{\mathrm{d}t} = cx_1 + dx_2 + b_2. \end{cases} \tag{1.252}$$

Writing $A_0 = \begin{pmatrix} a\ b \\ c\ d \end{pmatrix}$, $\boldsymbol{b} = {}^t(b_1, b_2)$ and $A = DE - A_0$, where $D = \frac{\mathrm{d}}{\mathrm{d}t}$ indicates the differential operator, we may write (1.252) as

$$A\boldsymbol{x} = \boldsymbol{b}. \tag{1.253}$$

Then it is natural to multiply the cofactor matrix $\tilde{A} = \begin{pmatrix} D-a & -b \\ -c & D-d \end{pmatrix}$ in operator entries. It follows that $\tilde{A}A = |A|E$, where $|A| = (D-a)(D-d) - bc = D^2 - (a+d)D + |A_0|$ is the determinant in operator entries and we have the pre-Cramer formula

$$\begin{cases} \frac{\mathrm{d}^2 x_1}{\mathrm{d}t^2} - (a+d)\frac{\mathrm{d}x_1}{\mathrm{d}t} + (ad-bc)x_1 = \frac{\mathrm{d}b_1}{\mathrm{d}t} - ab_1 - bb_2 \\ \frac{\mathrm{d}^2 x_2}{\mathrm{d}t^2} - (a+d)\frac{\mathrm{d}x_2}{\mathrm{d}t} + (ad-bc)x_2 = \frac{\mathrm{d}b_2}{\mathrm{d}t} - cb_1 - db_2. \end{cases} \tag{1.254}$$

Hence it suffices to solve one of them and substitute the solution into (1.252). Let $D = (a-d)^2 + 4bc$ and $\lambda_1 = \frac{a+d+\sqrt{D}}{2}$ with $\lambda_2 = \frac{a+d-\sqrt{D}}{2}$. Then if $\lambda_1 \neq \lambda_2$, we have

$$x_1 = c_1 e^{\lambda_1 t} + c_2 e^{\lambda_2 t} + x_{01}(t), \tag{1.255}$$

where x_{01} is a particular solution of the first equation in (1.254).

We now confine ourselves to the homogeneous system, i.e. we assume $\boldsymbol{b} = \boldsymbol{o}$. Then the derivative of (1.255) is $\frac{\mathrm{d}x_1}{\mathrm{d}t} = c_1\lambda_1 e^{\lambda_1} + c_2\lambda_2 e^{\lambda_2}$. Hence the initial values are $x_1(0) = c_1 + c_2$ and $ax_1(0) + bx_2(0) = x_1'(0) = c_1\lambda_1 + c_2\lambda_2$, so that

$$\begin{aligned} c_1 &= \frac{\lambda_2 x_1(0) - \lambda_1 x_1'(0)}{\lambda_2 - \lambda_1} = \frac{(\lambda_2 - a)x_1(0) - bx_2(0)}{\lambda_2 - \lambda_1}, \\ c_2 &= \frac{-(\lambda_1 - a)x_1(0) + bx_2(0)}{\lambda_2 - \lambda_1}. \end{aligned} \tag{1.256}$$

Hence

$$x_1(t) = \frac{1}{\lambda_2 - \lambda_1}\left(\left((\lambda_2 - a)e^{\lambda_1 t} - (\lambda_1 - a)e^{\lambda_2 t}\right)x_1(0) - b(e^{\lambda_1 t} - e^{\lambda_2 t})x_2(0)\right) \tag{1.257}$$

and

$$x_2(t) = \frac{1}{\lambda_2 - \lambda_1}\left(-c(e^{\lambda_1 t} - e^{\lambda_2 t})x_1(0) + \left((\lambda_2 - d)e^{\lambda_1 t} - (\lambda_1 - d)e^{\lambda_2 t}\right)x_2(0)\right). \tag{1.258}$$

The special case of $b = 0$ of (1.258) leads to the solutions in Example 5.3:

$$x_2(t) = e^{dt}x_2(0) + \frac{c}{a-d}\left(e^{at} - e^{dt}\right)x_1(0) \tag{1.259}$$

for $a \neq d$ because of $\lambda_1 = a, \lambda_2 = d$, which is (5.14).

Theorem 1.11. *Let $\boldsymbol{x}(t) = \begin{pmatrix} x_1(t) \\ x_2(t) \end{pmatrix}$ and let $A = \begin{pmatrix} a & b \\ c & d \end{pmatrix}$ be a matrix with eigenvalues λ_1, λ_2. The solutions of the homogeneous system of DEs*

$$\frac{\mathrm{d}}{\mathrm{d}x}\boldsymbol{x}(t) = A\boldsymbol{x}(t) \tag{1.260}$$

under the initial condition $\boldsymbol{x}(0) = \begin{pmatrix} x_1(0) \\ x_2(0) \end{pmatrix}$ *are given by linear combinations of* $e^{\lambda_1 t}$ *and* $e^{\lambda_2 t}$ *or* $te^{\lambda t}$ *and* $e^{\lambda t}$ *according to the eigenvalues that are different or the same,* $= \lambda$ *say. More concretely, in the latter case,*

$$\boldsymbol{x}(t) = P \begin{pmatrix} e^{\lambda t} & 0 \\ 0 & te^{\lambda t} \end{pmatrix} P^{-1} \boldsymbol{x}(0) \tag{1.261}$$

where

$$P = \begin{pmatrix} -\frac{2b}{a-d} & -\frac{4b}{(a-d)^2} \\ 1 & 0 \end{pmatrix}. \tag{1.262}$$

Proof. We consider the case where eigenvalues are $\lambda = \frac{a+d}{2}$ (double root) and the eigenspace is $E_A(\lambda) = \mathbb{C}\boldsymbol{p}_1$, $\dim E_A(\lambda) = 1$, the other cases being treated separately below. Here $\boldsymbol{p}_1 = \begin{pmatrix} p_{11} \\ p_{21} \end{pmatrix} = \begin{pmatrix} -\frac{2b}{a-d} \\ 1 \end{pmatrix}$ and the discriminant $D = (a-d)^2 + 4bc = 0$. Details are given in Exercise 13 below. We find $\boldsymbol{p}_2$ such that

$$P = (\boldsymbol{p}_1, \boldsymbol{p}_2) \in GL_2(\mathbb{C}) \tag{1.263}$$

entails

$$P^{-1}AP = \begin{pmatrix} \lambda & 1 \\ 0 & \lambda \end{pmatrix} = J(A), \tag{1.264}$$

the Jordan normal form. By block multiplication, we rewrite $AP = P\begin{pmatrix} \lambda & 1 \\ 0 & \lambda \end{pmatrix}$ into

$$\begin{aligned} (A\boldsymbol{p}_1, A\boldsymbol{p}_2) &= A(\boldsymbol{p}_1, \boldsymbol{p}_2) = AP = P\begin{pmatrix} \lambda & 1 \\ 0 & \lambda \end{pmatrix} \\ &= (\boldsymbol{p}_1, \boldsymbol{p}_2)\begin{pmatrix} \lambda & 1 \\ 0 & \lambda \end{pmatrix} = (\lambda\boldsymbol{p}_1, \boldsymbol{p}_1 + \lambda\boldsymbol{p}_2), \end{aligned} \tag{1.265}$$

whence we arrive at the system of equations

$$(A - \lambda E)\boldsymbol{p}_2 = \boldsymbol{p}_1. \tag{1.266}$$

Discarding the case where $a - \lambda = 0$, we may transform the augmented matrix as $(A - \lambda E, \boldsymbol{p}_1) \sim \begin{pmatrix} 1 & \frac{b}{a-\lambda} & \frac{p_{11}}{a-\lambda} \\ 0 & 0 & 0 \end{pmatrix}$ in view of $D = 0$ and so the solution is $\begin{pmatrix} \frac{p_{11}}{a-\lambda} \\ 0 \end{pmatrix} + c\begin{pmatrix} -\frac{b}{a-\lambda} \\ 1 \end{pmatrix}$, $c \in \mathbb{C}$. Putting $\boldsymbol{p}_2 = \begin{pmatrix} \frac{p_{11}}{a-\lambda} \\ 0 \end{pmatrix} = \begin{pmatrix} -\frac{4b}{(a-d)^2} \\ 0 \end{pmatrix}$, we find that $\{\boldsymbol{p}_1, \boldsymbol{p}_2\}$ are linearly independent. Hence putting P as in (1.263), we obtain (1.264).

We may now solve the DE (1.260). Putting

$$\boldsymbol{x} = P\boldsymbol{y}, \quad P^{-1}\boldsymbol{x} = \boldsymbol{y} = \begin{pmatrix} y_1 \\ y_2 \end{pmatrix}, \tag{1.267}$$

we see that (1.260) is equivalent to

$$\frac{\mathrm{d}}{\mathrm{d}t}\boldsymbol{y} = P^{-1}\frac{\mathrm{d}}{\mathrm{d}t}\boldsymbol{x} = P^{-1}AP\boldsymbol{y} = J(A)\boldsymbol{y}. \tag{1.268}$$

i.e. to

$$\begin{cases} \frac{\mathrm{d}y_1}{\mathrm{d}t} = \lambda y_1 + y_2 \\ \frac{\mathrm{d}y_2}{\mathrm{d}t} = \lambda y_2. \end{cases} \tag{1.269}$$

This is (5.13) with $x_1 = y_2$, $x_2 = y_1$, $k_1\ k_2 = -\lambda$.

Hence $y_1 = (y_2(0)t + y_1(0))\, e^{\lambda t}$ and

$$\boldsymbol{y} = \begin{pmatrix} e^{\lambda t} & 0 \\ 0 & te^{\lambda t} \end{pmatrix} \boldsymbol{y}(0) = \begin{pmatrix} e^{\lambda t} & 0 \\ 0 & te^{\lambda t} \end{pmatrix} P^{-1}\boldsymbol{x}(0). \tag{1.270}$$

Substituting (1.270) in (1.267), we prove (1.261). □

Exercise 13. Examine the case where the eigenvalue is a double root.

Solution. First assume $a = d$. Since the discriminant $D = 4bc$ is 0, we have either $b = 0$ or $c = 0$ and the matrix is either a lower or an upper triangular. We may therefore exclude this case.

In the case $a \neq d$ we have with the eigenvalue $\lambda = \frac{a+d}{2}$

$$A - \lambda E = \begin{pmatrix} \frac{a-d}{2} & b \\ c & -\frac{a-d}{2} \end{pmatrix} \sim \begin{pmatrix} 1 & \frac{2b}{a-d} \\ c & -\frac{a-d}{2} \end{pmatrix} \sim \begin{pmatrix} 1 & \frac{2b}{a-d} \\ 0 & 0 \end{pmatrix}$$

in view of $D = 0$. Hence $E_A(\lambda) = \left\{ c \begin{pmatrix} -\frac{2b}{a-d} \\ 1 \end{pmatrix} \middle| c \in \mathbb{C} \right\}$ and we may choose $\boldsymbol{p}_1$ as in the proof of Theorem 1.11.

We shall state the basics of diagonalizable case. This is another manifestation of exponential phenomenon.

Definition 1.7. For a square matrix A we define the exponential function in the matric argument by

$$e^A = \sum_{n=0}^{\infty} \frac{1}{n!} A^n. \tag{1.271}$$

Lemma 1.3. *If a square matrix A of degree k can be diagonalized by $P \in GL_k(\mathbb{C})$ as*

$$P^{-1}AP = \begin{pmatrix} \lambda_1 & & O \\ & \ddots & \\ O & & \lambda_k \end{pmatrix} = D, \tag{1.272}$$

then

$$A^n = P \begin{pmatrix} \lambda_1^n & & O \\ & \ddots & \\ O & & \lambda_k^n \end{pmatrix} P^{-1} \tag{1.273}$$

and

$$e^A = P \begin{pmatrix} e^{\lambda_1} & & O \\ & \ddots & \\ O & & e^{\lambda_k} \end{pmatrix} P^{-1}. \tag{1.274}$$

Proof. From (1.272), $A = PDP^{-1}$. Raising this to the nth power and noting the cancellation of PP^{-1}, we arrive at (1.273). Substituting (1.273) in (1.271), we deduce that

$$e^A = \sum_{n=0}^{\infty} \frac{1}{n!} P \begin{pmatrix} \lambda_1^n & & O \\ & \ddots & \\ O & & \lambda_k^n \end{pmatrix} P^{-1} = P \begin{pmatrix} \sum_{n=0}^{\infty} \frac{1}{n!}\lambda_1^n & & O \\ & \ddots & \\ O & & \sum_{n=0}^{\infty} \frac{1}{n!}\lambda_k^n \end{pmatrix} P^{-1}, \tag{1.275}$$

which leads to (1.274). □

In what follows we assume the coefficient matrix $A = (a_{ij})_{1\le i,j\le k}$ has constant entries. We consider the homogeneous system of DEs ($x_j = x_j(t)$)

$$\frac{\mathrm{d}}{\mathrm{d}t} \begin{pmatrix} x_1 \\ \vdots \\ x_k \end{pmatrix} = A \begin{pmatrix} x_1 \\ \vdots \\ x_k \end{pmatrix} \tag{1.276}$$

in the form

$$\frac{\mathrm{d}}{\mathrm{d}t}\boldsymbol{x}(t) = A\boldsymbol{x}(t), \tag{1.277}$$

a generalization of (1.260), by writing

$$\boldsymbol{x} = \boldsymbol{x}(t) = {}^t(x_1(t), \cdots, x_k(t)). \tag{1.278}$$

We may treat the linear DE of degree k

$$\frac{\mathrm{d}^k x}{\mathrm{d}t^k} + a_1 \frac{\mathrm{d}^{k-1} x}{\mathrm{d}t^{k-1}} + \cdots + a_{k-1} x = 0 \tag{1.279}$$

as a special case of (1.277) with

$$\boldsymbol{x} = \begin{pmatrix} x(t) \\ \vdots \\ x^{(k-1)}(t) \end{pmatrix} \tag{1.280}$$

and

$$A = \begin{pmatrix} 0 & 1 & 0 & \cdots & 0 \\ 0 & 0 & 1 & \cdots & 0 \\ & & \cdots & & \\ -a_{k-1} & & \cdots & & -a_1 \end{pmatrix}. \tag{1.281}$$

For differentiation of (1.280) gives $\frac{\mathrm{d}}{\mathrm{d}t}\boldsymbol{x}(t) = {}^t(x', \cdots, x^{(k)})$ and the last entry is $-a_{k-1}x - \cdots - a_1 \frac{\mathrm{d}^{k-1}x}{\mathrm{d}t^{k-1}}$ by (1.279) and so (1.279) amounts to (1.277).

Let $\boldsymbol{x}_n = \begin{pmatrix} x_1(n) \\ \vdots \\ x_k(n) \end{pmatrix}$. Then we may treat the system of difference equations (also called recurrence equations)

$$\Delta \begin{pmatrix} x_1 \\ \vdots \\ x_k \end{pmatrix} = A \begin{pmatrix} x_1 \\ \vdots \\ x_k \end{pmatrix}, \tag{1.282}$$

where Δ indicates the shift operator $\Delta x_l(n) = x_j(n+1)$, in the same way as we treat differential equations (the Pincherle duality) by writing

$$\Delta \boldsymbol{x}_n = \boldsymbol{x}_{n+1} = A\boldsymbol{x}_n. \tag{1.283}$$

We express the general term of the sequences $\{x_1(n)\}, \cdots, \{x_k(n)\}$ given by (1.283) in n.

Theorem 1.12. *Suppose A is a square matrix of degree k which is diagonalizable as in (1.272). Then the solution of the system of differential equations (1.277) under the initial condition $\boldsymbol{x}(0) = {}^t(x_1(0), \cdots, x_n(0))$ is given by*

$$\boldsymbol{x}(t) = e^{At}\boldsymbol{x}(0). \tag{1.284}$$

The solution of (1.283) under the initial condition $\boldsymbol{x}(0) = {}^t(x_1(0), \cdots, x_k(0))$ is given by

$$\boldsymbol{x}(k) = A^k \boldsymbol{x}(0). \tag{1.285}$$

Proof. By the change of variable (1.267) and (1.272), (1.277) amounts to

$$\frac{\mathrm{d}}{\mathrm{d}t}\boldsymbol{y}(t) = P^{-1}AP\boldsymbol{y}(t) = D\boldsymbol{y}(t)$$

correspondingly to (1.268), which in turn amounts to

$$\frac{\mathrm{d}}{\mathrm{d}t}y_j(t) = \lambda_j y_j(t). \tag{1.286}$$

Proposition 1.1 gives $y_j(t) = e^{\lambda_j t}y_j(0)$ or

$$\boldsymbol{y}(t) = \begin{pmatrix} e^{\lambda_1 t} & & O \\ & \ddots & \\ O & & e^{\lambda_k t} \end{pmatrix} \boldsymbol{y}(0). \tag{1.287}$$

Invoking (1.267) in the form $\boldsymbol{y} = P^{-1}\boldsymbol{x}$, we see that (1.287) means

$$\boldsymbol{x}(t) = P \begin{pmatrix} e^{\lambda_1 t} & & O \\ & \ddots & \\ O & & e^{\lambda_k t} \end{pmatrix} P^{-1}\boldsymbol{x}(0) \tag{1.288}$$

correspondingly to (1.261). The right-hand side is $e^{At}\boldsymbol{x}(0)$ in view of (1.274) and (1.284) follows.

The proof of (1.285) runs almost verbatim to the above proof. In place of (1.286) we have

$$\Delta y_j(n) = y_j(n+1) = \lambda_j y_j(n), \tag{1.289}$$

for which we have $y_j(n) = \lambda_j^n y_j(0)$. Hence correspondingly to (1.287), we have

$$\boldsymbol{y}(n) = \begin{pmatrix} \lambda_1^n & & O \\ & \ddots & \\ O & & \lambda_k^n \end{pmatrix} \boldsymbol{y}(0). \tag{1.290}$$

(1.273) now implies (1.285). □

1.7 Appendix B: Algebraic preliminaries

In this Appendix we shall give a very brief glimpse on algebraic structures used in §1.5.3 and Chapter 3. We shall confine to tensor product and alternating algebra. For this the basic is a free R-module, an additive Abelian group over a ring R which we view just as a generalization of the notion of a vector space over a field, e.g. over the familiar fields $\mathbb{R}$ or $\mathbb{C}$. By this we mean a free R-module has a basis in the sense similar to the basis of a vector space. We take the following proposition for granted.

Proposition 1.2. *With any given set as a basis we may construct a free module over a ring R.*

1.7.1 *Tensor product*

Let R be a ring with unity. For a right R-module M and a left R-module N, a mapping $f : M \times N \to T$, T being an Abelian group, is called a **semi-linear map** if it is additive with respect to both variables and $f(xa, y) = f(x, ay)$.

Definition 1.8. For a right R-module M and a left R-module N, their **tensor product** $M \otimes N = M \otimes_R N = (M \otimes N, \otimes)$ is an Abelian group satisfying the following conditions. The map

$$\otimes : M \times N \to M \otimes N, (x, y) \mapsto x \otimes y, \quad x \in M, y \in N \tag{1.291}$$

satisfies the semi-bilinearity conditions

$$\begin{aligned}(x_1 + x_2, y) &= (x_1, y) + (x_2, y) \\ (x, y_1 + y_2) &= (x, y_1) + (x, y_2) \\ (xa, y) &= (x, ay)\end{aligned} \tag{1.292}$$

and also universality condition (U) is satisfied:
(U) For any semi-bilinear map $\tau : M \times N \to T$ to an Abelian group, there exists a unique $t : M \otimes N \to T$ such that $\tau = t \circ \otimes$.

Theorem 1.13. *The tensor product $M \otimes_R N$ exists and is unique up to a isomorphism. Its elements are expressed as a finite linear combination*

$$\sum x_i \otimes y_i, (x_i \in M, y_i \in N). \tag{1.293}$$

Proof. Existence.
Let $\mathcal{F} = \mathcal{F}_{M\times N}$ be a free Abelian group i.e. a free $\mathbb{Z}$-module with basis $M \times N$. The relation $\sim$ is to be set so as to make the reduction of a system satisfy the conditions for a tensor product, i.e. semi-bilinearity of the natural projection $\pi_\sim$. I.e. for $x, x_i \in M, y, y_j \in N, a \in R$, the reduction $\sim$ is to make (1.292) hold true in $\mathcal{F}/\sim$. I.e. two elements are equivalent if they differ by elements in (1.294), or simply their difference belongs to the subgroup $\mathcal{H} \subset \mathcal{F}$ generated by elements of the form

$$\begin{aligned}&(x_1 + x_2, y) - (x_1, y) - (x_2, y) \\ &(x, y_1 + y_2) - (x, y_1) - (x, y_2) \\ &(xa, y) - (x, ay).\end{aligned} \tag{1.294}$$

Then $\mathcal{S} = \mathcal{F}/\sim = \mathcal{F}/\mathcal{H}$ as a factor group. The canonical projection is $\pi = \pi_\sim : \mathcal{F} \to \mathcal{S}$. The restriction $\pi_{M\times N}$ is a semi-bilinear map and works as the map $\otimes$ in the definition. Any semi-bilinear map $\tau : M \times N \to T$ can be extended to a homomorphism $\mathcal{F} \to T$ since it maps $\mathcal{H}$ to 0. Hence we may define the map $t : \mathcal{F}/\mathcal{H} \to T$ by $t \circ \pi_\sim = \tau$. Hence $\mathcal{F}/\mathcal{H}$ satisfies the defining conditions in Definition 1.8 and passes as the tensor product $M \otimes N$.

Uniqueness.
Suppose in addition to $(M \otimes N, \otimes)$, the pair (T, τ) also satisfies the universality condition (U) is satisfied, i.e. there exists a $t_1 : T \to M \otimes N$ such that

$$(t_1 \circ t)(x \otimes y) = (t_1 \circ \tau)(x, y) = x \otimes y.$$

Applying the uniqueness in (U) to $T = M \otimes N$ and $\tau = \otimes$, we see that $t_1 \circ t = I_{M\otimes N}$. Changing the role, we have $t \circ t_1 = I_T$. Hence $t : M \otimes N \simeq T$. Identifying $t : M \otimes N$ and T, the maps $\otimes$ and T are identified and the uniqueness holds including the semi-bilinear map (1.291).

Linear combination expression.
Let T be the set of all elements of the form (1.293) forms an Abelian group. The map $\otimes$ in (1.291) induces a semi-bilinear map $M \times N \to T$. Universality implies the existence of a homomorphism $t : M \otimes N \to T (\subset M \otimes N)$ such that $t(x \otimes y) = x \otimes y$. By the above argument, this t must be an isomorpshism and $T = M \otimes N$. □

Theorem 1.14. *If M is a free R-module with basis $\{u_\mu\}$, then the elements of $M \otimes_R N$ can be expressed uniquely in the form*

$$\sum u_\mu \otimes n_\mu, \tag{1.295}$$

where $n_\mu \in N$ all but finite number of them being 0.

1.7.2 *Graded algebras*

Definition 1.9. Let J be a commutative semi-group with unity 0, we shall confine to $J = \{0, 1, 2, \cdots\}$. The direct sum $M = \sum_{j\in J} M_j$ of R-modules M_j with index set J is called a graded R-module of type J. Every element in M_j is called a **homogeneous element** of degree j. Any submodule $N \subset M$ of the form $\sum_{j\in J} N_j$, $N_j \subset M_j$ is called a graded R-submodules.

If in a graded module $A = \sum A_k$ over a commutative ring R the product is defined so that $A_k A_l \subset A_{k+l}$ and A forms an algebra with respect to this product, then A is called a **graded algebra**.

Definition 1.10. Let R be a commutative ring with unity and let M be an R-module. Let $T^k(M)$ be the k-ple tensor product

$$T^k(M) = M \otimes \cdots \otimes M. \tag{1.296}$$

There is a natural isomorphism $T^k(M) \otimes T^l(M) \simeq T^{k+l}(M)$. With $T^0(M) = R$ form the graded module

$$T(M) = \sum_k T^k(M) = R \otimes (M) \otimes (M \otimes M) \otimes (M \times M \times M) \otimes \cdots . \tag{1.297}$$

The product of $x_1 \otimes \cdots \otimes x_k \in T^k(M)$ and $y_1 \otimes \cdots \otimes y_l \in T^l(M)$ is defined by

$$(x_1 \otimes \cdots \otimes x_k) \otimes (y_1 \otimes \cdots \otimes y_1) = x_1 \otimes \cdots \otimes x_k \otimes y_1 \otimes \cdots \otimes y_l \in T^{k+l}(M) \tag{1.298}$$

according to the above natural isomorphism. (1.297) becomes a graded algebra, called the **tensor algebra** or a tensor product algebra.

Definition 1.11. A k-ple linear map $f : M^k \to N$ is called **alternating** if it satisfies one of the following equivalent conditions
(i) If $x_i \neq x_j$ for $i \neq j$ then $f(x_1, \cdots, x_k) = 0$.
(ii) If x_i, x_j with $i \neq j$ are changed, it changes sign: $f(x_1, \cdots, x_j, \cdots, x_i, \cdots x_k) = -f(x_1, \cdots, x_i, \cdots, x_j, \cdots x_k)$.

In $T^k(M)$ we introduce a relation: $x \sim y$ if they differ by elements $x_1 \otimes \cdots \otimes x_k$ in which $x_i = x_j$ for some $i \neq j$. Or we let $\mathfrak{a}_k$ be the submodule consisting of those elements $x_1 \otimes \cdots \otimes x_k$ in which at least two components coincide and we define the R-module

$$\Lambda^k(M) = T^k(M)/\sim = T^k(M)/\mathfrak{a}_k; \quad \pi_\sim : T^k(M) \to \Lambda^k \tag{1.299}$$

and write

$$\pi_\sim(x_1 \otimes \cdots \otimes x_k) = x_1 \wedge \cdots \wedge x_k. \tag{1.300}$$

Hence $x_1 \wedge \cdots \wedge x_k$ are alternating.

With $\mathfrak{a}_0 = 0$, the ideal $\mathfrak{a} = \mathfrak{a}_M = \sum_k \mathfrak{a}_k$ forms a homogeneous ideal of $T(M)$ and is generated by all $x \otimes x$, so that

$$\Lambda(M) := T(M)/\mathfrak{a} = \sum_{k=0}^{\infty} T^k(M)/\mathfrak{a}_k \tag{1.301}$$

is a graded module whose elements are alternating.

Theorem 1.15. *Every linear map $f : M \to N$ from an R-module M to another R-module N is extended uniquely to an algebra homomorphism*

$$T(f) : T(M) \to N. \tag{1.302}$$

If N is an algebra, then

$$T(f)(x_1 \otimes \cdots \otimes x_k) = f(x_1) \cdots f(x_k). \tag{1.303}$$

Proof. Since $T(M)$ is a graded module, it suffices to consider its $(k+1)$th component $T^k(M)$ on which we should have

$$T(f)(x_1 \otimes \cdots \otimes x_k) = f(x_1, \cdots, x_k), \qquad (1.304)$$

whence $T(f)$ is unique. The left-hand side of (1.304) is to be the right-hand side of (1.303) if N is an algebra. Since (1.303) is a linear map and $T(f)$ exists. Moreover, by the definition of the product in N, it is an algebra homomorphism. □

modules	linear map, alg homom
(0) $N = T(N)$ (tensor algebra)	f
(i) $N = A$ (algebra)	$f(x_1, \cdots, x_k) = f(x_1) \cdots f(x_k)$
(ii) $M = M^k$	$f(x_1, \cdots, x_k)$
(iii) $M = M_1 \oplus M_2, N = \Lambda(M_1) \otimes \Lambda(M_2)$	f in (1.306)

Table 1.3: Special cases

Corollary 1.2. (i) *Every linear map $f : M \to N$ is extended to a graded algebra homomoprphism $f : T(M) \to T(N)$. This induces an algebra homomorphism*

$$\Lambda(f) : \Lambda(M) \to \Lambda(N)$$

given by

$$\Lambda(f)(x_1 \wedge \cdots \wedge x_n) = f(x_1) \wedge \cdots \wedge f(x_n).$$

(ii) *A k-ple linear map $f : M^k \to N$ viewed as a linear map $T(f) : T^k(M) \to N$ is alternating if and only if there exists a linear map $\bar{T}(f) : T^k(M) \to N$ such that*

$$f(x_1, \cdots, x_k) = \bar{T}(f)(x_1 \wedge \cdots \wedge x_k). \qquad (1.305)$$

(iii) *The linear map $f : M_1 \oplus M_2 \to \Lambda(M_1) \otimes \Lambda(M_2)$ given by*

$$f(x + y) = x \otimes 1 + 1 \otimes y \qquad (1.306)$$

is extended to a graded algebra homomoprphism $T(f) : T(M_1 \oplus M_2) \to \Lambda(M_1) \otimes \Lambda(M_2)$, which in turn induces a graded algebra homomoprphism $\bar{T}(f) : \Lambda(M_1 \oplus M_2) \to \Lambda(M_1) \otimes \Lambda(M_2)$ satisfying

$$\bar{T}(f)(x_1 \wedge \cdots \wedge x_k \wedge y_1 \wedge \cdots \wedge y_l) = (x_1 \wedge \cdots \wedge x_k) \otimes (y_1 \wedge \cdots \wedge y_l), x_i \in M_1, y_j \in M_2. \qquad (1.307)$$

Proof. The cases correspond to (i)-(iii) in Table 1.3. (i) follows from Theorem 1.15 on viewing $f : M \to N$ as $f : M \to T(N)$. The second assertion follows since $T(f)$ maps $\mathfrak{a}_M$ to $\mathfrak{a}_N$.
(ii) follows from universality. By Theorem 1.15 we may view f as a linear map $T(f) : T^k(M) \to N$. Universality means $T(f) = \bar{T}(f)(\pi_\sim)$, which implies (1.305) in view of (1.300).
(iii) By Theorem 1.15, f is extended to $T(f)$. By (1.303), we have for $z = x + y \in M_1 \oplus M_2$, $x \in M_1$, $y \in M_2$

$$\begin{aligned} T(f)(z \otimes z) &= f(z)f(z) = (x \otimes 1 + 1 \otimes y)(x \otimes 1 + 1 \otimes y) \\ &= (x \otimes 1)(1 \otimes y) + (1 \otimes y)(x \otimes 1) = 0 \end{aligned}$$

by alternating property of the product in $\Lambda(M_1) \otimes \Lambda(M_2)$. Hence by universality, $T(f)$ induces $\bar{T}(f)$ satisfying (1.307). □

Lemma 1.4.

$$\Lambda(M_1 \oplus M_2) \simeq M_1 \otimes M_2. \tag{1.308}$$

Proof. The injection

$$\iota_j : M_j \to M_1 \oplus M_2, \quad j = 1, 2$$

induces the algebra homomorphism

$$\Lambda(\iota_j) : \Lambda(M_j) \to \Lambda(M_1 \oplus M_2).$$

Hence the algebra homomorpshim $g : \Lambda(M_j) \otimes \Lambda(M_2) \to \Lambda(M_1 \oplus M_2)$ is induced such that

$$g((x_1 \wedge \cdots \wedge x_k) \otimes (y_1 \wedge \cdots \wedge y_l)) = (x_1 \wedge \cdots \wedge x_k) \wedge (y_1 \wedge \cdots \wedge y_l).$$

Since this is the inverse map of $\bar{T}(f)$, the isomorphism (1.308) is given by $\bar{T}(f)$. □

Theorem 1.16. *Suppose M is a free R-module with basis $\{u_1, \cdots, u_n\}$: $M = Ru_1 \oplus \cdots \oplus Ru_n$. Then for $k > n$, $\Lambda^k(M) = 0$ and for $k \leq n$, $\Lambda^k(M)$ is a free R-module of rank $\binom{n}{k}$ with basis $\{u_{i_1}, \cdots, u_{i_k}\}$, $i_1 < \cdots < i_k$. $\Lambda^n(M)$ is a free R-module of rank 2^n.*

Proof. By Lemma 1.4,

$$\Lambda(M) \simeq \Lambda(Ru_1) \otimes \cdots \otimes \Lambda(Ru_n).$$

Since $\Lambda(Ru_j) = R1 \oplus Ru_j$, the result follows. □

Remark 1.5. The differential forms in Chapter 3 can be concisely introduced by the above. Let $R = C^r(\Omega)$ and $M = Ru_1 \oplus \cdots \oplus Ru_n$ be a free R-module of rank n. Then the k-ple tensor product $T^k(M)$ is $\{\omega^k = \sum_I a_I u_I\}$ with $I = \{i_1, \cdots, i_k\} \subset \{1, \cdots, n\}$. With $\mathfrak{a}_k$ as above and $\mathrm{d}x_j$, the space of k-forms is

$$\Lambda^k(M) = T^k(M)/\mathfrak{a}_k = \left\{ \sum_I a_I \mathrm{d}x_I \right\}, \tag{1.309}$$

where $\mathrm{d}x_j$'s have alternating properties.

Fig. 1.5: Gaspard-Gustave Coriolis 1792-1843

Chapter 2

Thermodynamics in engine principles

Abstract

When there are much less fossil fuels available, one of the main issues will be reducing the friction loss. For this, lubrication is a solution which depends on fluid mechanics. We shall elucidate the origin of the governing principle—the Navier-Stokes equation in its full generality for compressible fluid. The argument is rather transparent in that we start from the Cauchy equation of motion which is a counterpart of the Newton equation of motion in fluid mechanics and derive the Navier-Stokes equation via a result of Stokes, showing that fluid mechanics is a version of mechanics. In the beginning, we give some basics of vector analysis and mathematical foundations of fluid dynamics.

2.1 The Boyle-Charles law

Let P, V and T denote the **pressure**, **volume** and **temperature** of an object (in our case it is a gas), respectively. Here the unit of T must be treated with care because of the following **Charles' law**:
Let ϑ denote the temperature of a gas in Celsius (°C) and let V_0 be its volume at time 0. Then under constant pressure, the volume V at time t is given by

$$V = \left(1 + \frac{1}{273.15}\right) V_0, \tag{2.1}$$

i.e. under constant pressure, for an increase 1 (°C) of a gas, its volume increases by $\frac{1}{273.15}$ times its initial volume at time 0.

It is convenient to introduce the new unit of temperature, **absolute temperature**, in K (Kelvin) introduced by Lord Kelvin (neé Thomson) and denoted by T. Between the two unit systems Celsius and absolute temperature, the relation

$$T = \vartheta + 273.15 \tag{2.2}$$

holds. In view of (2.2), the Charles' law (2.1) may be written as

$$V = \frac{T}{273.15} V_0. \tag{2.3}$$

Two special cases of (2.2) are noteworthy:

$$0\,(K) = -273.15\,(^\circ\mathrm{C}), \tag{2.4}$$

which is called the **absolute zero** and there cannot exist temperature lower than this;

$$273.15\,(K) = 0\,(^\circ\mathrm{C}).$$

There is another governing law, the **Boyle's law** which says that under constant temperature, pressure and volume (of a gas) are inversely proportional:

$$PV = \text{constant}. \tag{2.5}$$

In thermodynamics, such gases that satisfy Boyle's law are called **ideal gases**. Examples of ideal gases are Helium He, hydrogen H_2, oxygen O_2, nitrogen N_2 etc., which are not easily liquidated.

Given n mol ideal gas, the **Boyle-Charles law**

$$PV = nRT \tag{2.6}$$

holds, where R is the **gas constant** given by

$$R = 8.31441\ (\mathrm{Jmol^{-1}K^{-1}}). \tag{2.7}$$

(2.6) is called the **equation of state** of the ideal gas.

Let m and M denote the mass and molecular weight of the gas, then $n = \frac{m}{M}$ and (2.6) may also be written as

$$PV = \frac{m}{M} RT.$$

A more advanced state equation than (2.6) is the **van der Waals state equation**

$$\left(P + \frac{an^2}{V^2}\right)(V - bn) = nRT, \tag{2.8}$$

where $a > 0$ and $b > 0$ are constants intrinsic to the gas.

2.2 Engine terminology

We assemble here some of the most important terminologies on the engine.

- V_c designates the **clearance volume**, which is the volume of the cylinder when the piston is at T.D.C = Top Dead Center (or T.D.P. = Top Death Point). This is the volume which always remains in the cylinder, i.e. the mixed gas cannot be compressed more than this.

- L means the **stroke**, which is the length of the movement of the piston from T.D.C. to B.D.C. (= Bottom Death Center).

- V_h means the **stroke volume** or the **displacement volume**, which is the cylinder volume corresponding to one stroke = $\pi\left(\frac{D}{2}\right)^2 L$, where D is the diameter of the cylinder.

- $V_t = V_c + V_h$ is the **cylinder volume**.

V_h is used to indicate the power of the engine, like a car of 2000 cc meaning one with exhaust volume 2000 cc. T.D.C. and B.D.C. mean the position at which the piston can reach the highest point (T.D.P.) and the lowest point (B.D.P.) inside the cylinder.

The **compression ratio** ε is defined by

$$\varepsilon = \frac{V_t}{V_c}. \tag{2.9}$$

We have $\varepsilon = 1 + \frac{V_h}{V_c}$ or $V_c = \frac{1}{\varepsilon - 1} V_h$.

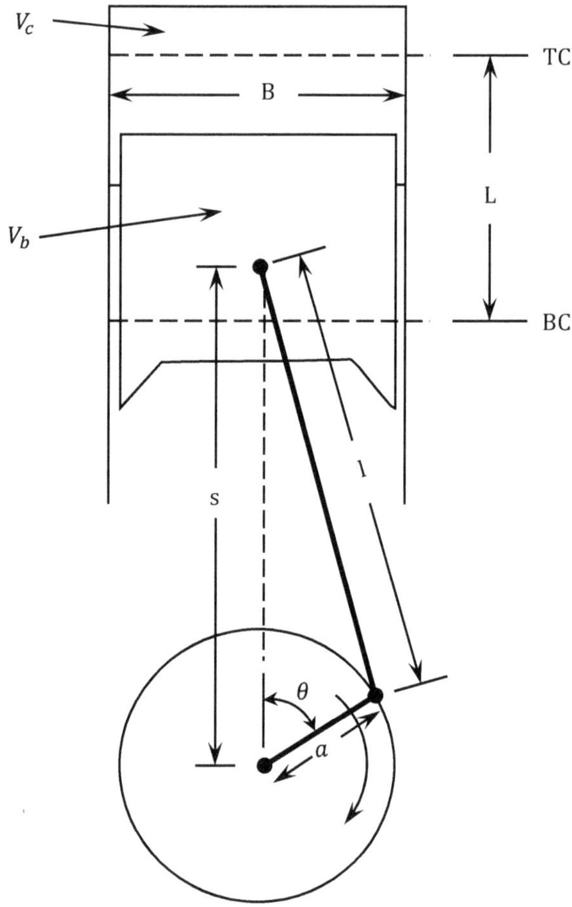

Fig. 2.1: Cylinder and piston

Example 2.1. Given a one-cylinder engine with cylinder diameter $D = 7$ cm, stroke $L = 8.2$ cm, clearance volume $V_C = 57.5$ cc, we find the stroke volume and the compression ratio. Find the exhaust volume of the 4-cylinder engine with the same spec.

Indeed, we have $V_h = \pi\left(\frac{7}{2}\right)^2 8.2 \approx 316$. $V_t = V_c + V_h \approx 373.5$ and $\varepsilon = \frac{373.5}{57.5} = 6.5$. In the case of a 4-cylinder engine, the total exhaust volume is $V_s = 4V_h \approx 1264$ cc.

In the case of an Otto cycle, ε remains within the range 9$\sim$10 while in the case of a Diesel engine, ε can be made as large as 18∼23 so that by the formula for the efficiency to be stated below, the heat efficiency of a Diesel cycle is better than that of an Otto cycle.

2.2.1 *Combustion processes*

The combustion processes of an Otto cycle and that of a Diesel cycle are mostly different except for an initial period of premixed combustion. We start from the combustion of an Otto cycle.

(i) Premixed Combustion.

The combustion of a gasoline engine (Otto cycle) is premixed combustion in which the mixture of the air and the almost **homogeneous fuel gas** created by the carburetor, being ignited by the spark made by the ignition plug, burns (in the case of a practical engine, in the mixture, there is the remaining gas contained from the previous combustion). The combustion process is the **flame propagation** in which the **flame front** of 1 mm width propagates which is really burning. In the engine chamber there is a micro-turbulence occurring, so that the propagation speed reaches 50 m/s at 6000 rpm. But heat receiving is not done in an instant.

(ii) Diffusion Combustion.

The combustion of a Diesel engine is that of diffusion type, i.e. only **natural air** is absorbed during the suction process, which is compressed under heavy pressure amounting to 30–45 kg/cm^2 (gauge) to reach a **high temperature** 500°–600°. Here the compression ratio ε defined by (2.6) reaches 22–30, which is much bigger compared to that of an Otto cycle, which remains within 9–10. Into this high temperature air, the **atomized light oil** is injected whose self-ignition temperature is 300°–400°.

(i-1) In an ignition lag of 0.007–0.030 sec., the injected fuel being heated, gets vaporized to make **mixed gas**. Since the mixed gas is **not homogeneous**, there are many points formed in the chamber that reach the ignition temperature, called **flame cores**.

(i-2) Flame propagation. The mixed gas near the nozzle gets self-ignited first and by flame propagation they propagate to many flame cores and the whole mixed gas gets ignited in the whole chamber and the combustion takes place almost in an instant. This period is similar to that of an Otto cycle, but the pressure-rising rate is rather high, makes noise and because of the high temperature, it creates plenty of NO_x. Also, during this period, the **diffusion of the air cannot catch up to the speed of**

thermal dissociation of the fuel, there arise plenty of black smoke (dry soot). Although most of this soot is burned during the successive diffusion combustion period, the remainder is exhausted into the air.
(ii-1) (When the engine load is high) the remaining vapor penetrates the flame and burns by absorbing the surrounding air. This process is called a **diffusion combustion**. In this combustion, the flame front is not thick and the flame belt exits only on the surface where the density of the mixed air is dense enough to approximate the theoretical rate of the mixed air. Inside the belt, the air/fuel ratio is small, so that in view of the deficit of oxygen, dry soot tends to arise.

We are to pay attention to the first passage of (ii), i.e. the explanation of the suction stroke of a Diesel cycle. It says without reasoning that the air is compressed to 1/22–1/30 times to reach high temperature of 500°–600°. We note here the Boyle-Charles law that applies to give the sound reasoning why the temperature goes up. I.e., although air is compressed, the volume cannot become too small compared with the pressure which is 30–45 kg/cm^2, so that the temperature, being proportional to the product of pressure and volume, which becomes big, must goes up.

2.3 The first law of thermodynamics

In thermodynamics, the distinction of two kinds of variables are made, i.e. the extensive variable and intensive variable, where the former depends on the change of the object while the latter does not change, e.g. the pressure remains the same during changes and it is often denoted by p rather than P. But we shall not adopt this distinction in most cases. **The first law of thermodynamics** or the **law of conservation of energy** is one of the most universal laws that governs our space. In addition to the symbols P, V, T introduced in §2.1, we need a few more to describe the basics of thermodynamics. We consider an isolated thermodynamical system, where isolated means that the system does not give or receive heat from outside sources. Although we are to use p to denote the state quantity, we follow the notation commonly used.

- U means the **inner energy**

- Q means the **heat**

- W means the **work**
- $S = \frac{Q}{T}$ means the **entropy**

The last is a temporary definition of entropy and a more proper definition is given by

$$\mathrm{d}S = \frac{\mathrm{d}Q_{rev}}{T} + \mathrm{d}S_{irr}, \tag{2.10}$$

where $\mathrm{d}Q_{rev}$ is the infinitesimal change of the heat under a reversible process and $\mathrm{d}S_irr$ is that of the entropy under an irreversible process. Also, to be most precise, we need to place a conversion factor = **mechanical equivalent to heat**,

$$A = 1/426.8 \text{ kcal/kg} \cdot \text{m}, \tag{2.11}$$

to convert the heat in work:

$$AW = Q_2 - Q_1, \tag{2.12}$$

where Q_2 and Q_1 are the heat exerted on and emitted from the system and W is the work done on the system. However, we may discard this factor if we measure the heat in Joules. Or if we take the ratio, such as heat efficiency, this disappears.

As we have the equation of state (2.2), we always have a relation between these variables and the situation is such that there are only two variables can be independent. The first law of thermodynamics then reads

$$\Delta U = \Delta Q + \Delta W, \tag{2.13}$$

i.e. the sum of the energy exerted on an object, to change its state, in the form of work W and of heat Q is independent of the way they are exerted.

This implies the negation of the existence of a perpetual engine of the first kind, i.e.

"Realizing a system which repeats a cyclic movement (e.g. rotation) without exerting energy from outside is impossible."

Now for the ideal gas, the **Joule law** is known to hold:

$$\Delta U = mc_V \Delta T, \tag{2.14}$$

where c_V is the constant volume specific heat (of the gas). Suppose the process is adiabatic, i.e. $\Delta Q = 0$. Substituting (2.14) with $m = 1$ in (2.13), we get

$$c_V \Delta T + P \Delta V = 0. \tag{2.15}$$

Dividing (2.15) by (2.4) in the form $P\frac{V}{m} = rT$, where $r = \frac{R}{M}$ and passing to the limit, we obtain

$$0 = c_V \frac{\mathrm{d}T}{T} + r\frac{\mathrm{d}V}{V} = c_V \frac{\mathrm{d}T}{T} + (c_P - c_V)\frac{\mathrm{d}V}{V},$$

where we used **Mayer's law**

$$r = c_P - c_V \left(= A\frac{R}{M}\right), \tag{2.16}$$

with M designating the molecular weight of the gas.

Recalling the **specific heat ratio**

$$k = \frac{c_P}{c_V}, \tag{2.17}$$

we transform the above equality into

$$\frac{\mathrm{d}T}{T} + (k-1)\frac{\mathrm{d}V}{V} = 0,$$

whence

$$\log T + (k-1)\log V = C \,(\text{constant}),$$

or

$$TV^{k-1} = \text{constant}. \tag{2.18}$$

We now apply (2.18) to the compression stroke ($1 \to 2$) of an engine (cf. §2.5), which is assumed to be adiabatic. Let V_1, V_2 and $T_1.T_2$ denote the volume and temperature in the beginning and the end of the stroke. Then (2.18) means that $T_2V_2^{k-1} = T_1V_1^{k-1}$ or using (2.13)

$$\frac{T_2}{T_1} = \left(\frac{V_1}{V_2}\right)^{k-1} = \varepsilon^{k-1}.$$

This may be also written

$$\frac{P_2}{P_1} = \varepsilon^k.$$

Hence we have **mathematically** deduced the equation of change of pressure and volume in the compression process (which is often described as experimental laws):

$$P_2 = \varepsilon^k P_1, \quad T_2 = \varepsilon^{k-1}T_1. \tag{2.19}$$

2.4 Thermodynamic intermission

Notation:

$\left(\frac{\partial S}{\partial T}\right)_V =$ the partial derivative of S as a function in T, V with respect to T with V viewed as a constant in T.

Example 2.2. The state of an object in thermal equilibrium can be described by the variables T, S, P, U. We have the relation

$$\left(\frac{\partial S}{\partial T}\right)_V \left(\frac{\partial P}{\partial V}\right)_S = \left(\frac{\partial S}{\partial T}\right)_p \left(\frac{\partial P}{\partial V}\right)_T. \tag{2.20}$$

This can be easily proved if we view the partial derivative as a Jacobian, e.g.

$$\left(\frac{\partial T}{\partial P}\right)_V = \frac{\partial(T,V)}{\partial(P,V)} = \frac{\partial(V,T)}{\partial(V,P)} = -\frac{\partial(V,T)}{\partial(P,V)}. \tag{2.21}$$

By (2.21), the left-hand side of (2.20) is

$$-\frac{\partial(S,V)}{\partial(T,V)}\frac{\partial(P,S)}{\partial(S,V)},$$

which is $-\frac{\partial(P,S)}{\partial(T,V)}$ by the multiplication theorem for the Jacobians. This last expression may be written as

$$-\frac{\partial(S,P)}{\partial(V,T)} = \frac{\partial(S,P)}{\partial(T,P)}\frac{\partial(P,T)}{\partial(V,T)},$$

which is the right-hand side of (2.20).

Definition 2.1. A function in a closed thermodynamical system is a state variable if and only if it can be expressed as an exact differential. We illustrate this by the interior energy $U = U(S,V)$:

$$\mathrm{d}U = \left(\frac{\partial U}{\partial S}\right)_V \mathrm{d}S + \left(\frac{\partial U}{\partial V}\right)_S \mathrm{d}V. \tag{2.22}$$

Example 2.3. The fundamental equation in thermodynamics is

$$\mathrm{d}U = T\mathrm{d}S - P\mathrm{d}V \tag{2.23}$$

and the **Maxwell equation**

$$\left(\frac{\partial T}{\partial V}\right)_S = -\left(\frac{\partial P}{\partial S}\right)_V \tag{2.24}$$

holds.

Indeed, comparing (2.22) and (2.23), we have

$$T = \left(\frac{\partial U}{\partial S}\right)_V, \quad -P = \left(\frac{\partial U}{\partial V}\right)_S.$$

Hence the left-hand side of (2.24) is

$$\left(\frac{\partial}{\partial V}\left(\frac{\partial U}{\partial S}\right)_V\right)_S,$$

which is, by the **Schwarz theorem**,

$$\left(\frac{\partial}{\partial S}\left(\frac{\partial U}{\partial V}\right)_S\right)_V.$$

This last expression is $\left(\frac{\partial}{\partial S}(-P)\right)_V$ = right-hand side of (2.24).

Exercise 14. Let H denote the **enthalpy** of the system defined by

$$H = U + PV. \tag{2.25}$$

Use (2.23) to deduce that

$$\mathrm{d}H = T\mathrm{d}S + V\,\mathrm{d}P \tag{2.26}$$

and that

$$\left(\frac{\partial T}{\partial P}\right)_T = -\left(\frac{\partial V}{\partial S}\right)_P.$$

Solution. Substituting (2.23) in

$$\mathrm{d}H = \mathrm{d}U + P\,\mathrm{d}V + V\,\mathrm{d}P,$$

(2.30) immediately follows. Since the enthalpy is a state quantity, we have

$$T = \left(\frac{\partial H}{\partial S}\right)_P, \quad V = \left(\frac{\partial H}{\partial P}\right)_S.$$

The remaining part is the same as in Example 2.3.

Example 2.4. Let F denote the **Helmholtz free energy** of the system defined by

$$F = U - TS. \tag{2.27}$$

We use (2.23) to deduce that

$$\mathrm{d}F = -S\mathrm{d}T - P\,\mathrm{d}V \tag{2.28}$$

and that

$$\left(\frac{\partial S}{\partial V}\right)_T = \left(\frac{\partial P}{\partial T}\right)_V.$$

Example 2.5. Let G denote the **Gibbs free energy** of the system defined by

$$G = U - TS + PV. \tag{2.29}$$

We may deduce that

$$\mathrm{d}G = -S\mathrm{d}T + V\,\mathrm{d}P \tag{2.30}$$

and that

$$\left(\frac{\partial S}{\partial P}\right)_T = -\left(\frac{\partial V}{\partial T}\right)_P.$$

2.5 Engine strokes

Inner combustion engines (Sabathé cycles) repeat the cycle consisting of four strokes:

suction (5→1) → compression (1→2) → expansion (2→3,4) → exhaustion (4→5),

where compression and expansion strokes are governed by the Boyle-Charles law in the sense that in the former, the (drastic) increase in P gives rise to, in view of the presence of the clearance volume, the (drastic) increase in T, while in the latter, the (drastic) increase in T gives rise to the (drastic) decrease in P by which V increases to V_h to make work on the piston which is forced to go down to B.D.C.

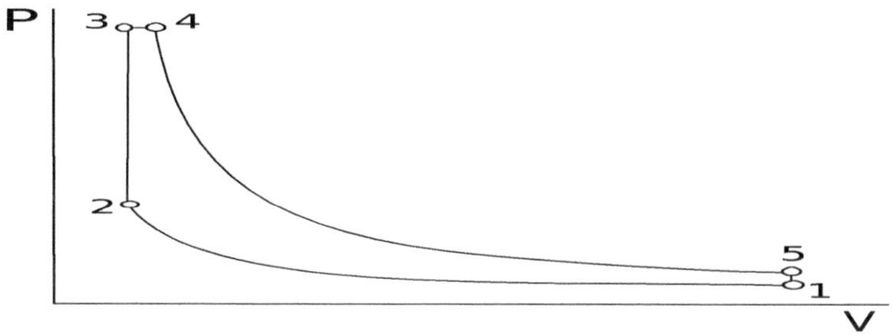

Fig. 2.2: PV plot for a Sabathé cycle

Theoretically, the compression stroke (1→2) and the expansion stroke (2→3,4) are assumed to be **adiabatic**. This explains why the PV graph is of the form of a (orthogonal) hyperbola due to the Boyle-Charles law again. The expansion stroke is preceded by heat receiving process which consists of two different processes,

(1) constant volume heat receiving (combustion) (2→3) = Otto cycle

(2) constant pressure heat receiving (combustion) (3→4) = Diesel cycle.

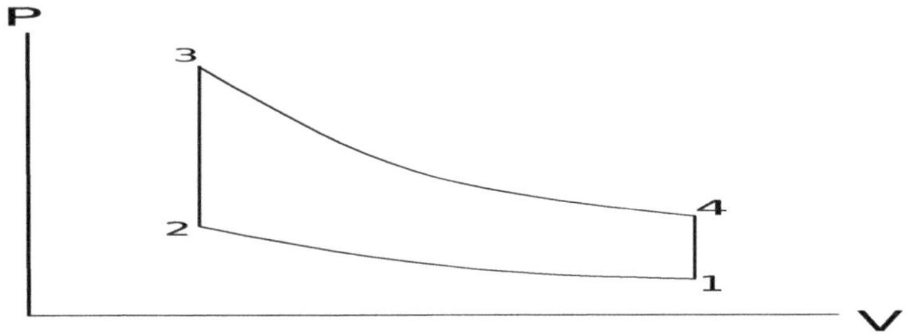

Fig. 2.3: PV plot for an Otto cycle

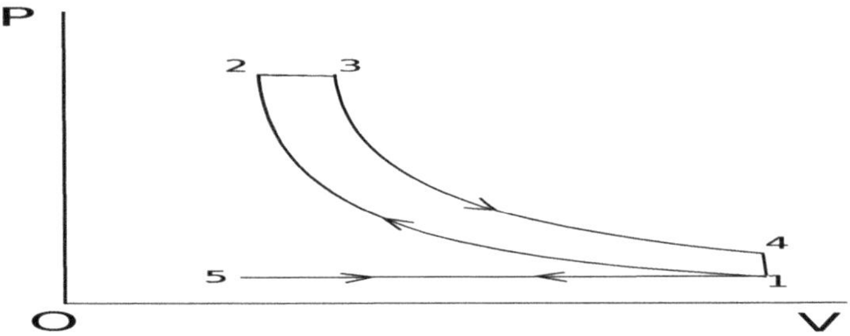

Fig. 2.4: PV plot for a Diesel cycle

In both cases, by the first law of thermodynamics (2.13), the increase in the interior energy is given as the increase in heat received:

$$\Delta U_V = \Delta Q_V = mc_V (T_3 - T_2), \Delta U_P = \Delta Q_P = mc_P (T_4 - T_3), \quad (2.31)$$

where m is the mass of the gas (kg), c_V and c_P are (constant volume) specific heat and (constant pressure) specific heat of the gas, respectively defined in §2.3. After this "instantaneous heat receiving", expansion starts and pushes down the piston. Since we assume the process to be adiabatic, we have (2.18). In order to deduce counterparts of (2.19), we introduce two more ratios. **Explosion ratio** is defined by

$$\sigma = \frac{P_3}{P_2}, \quad (2.32)$$

while **cut-off ratio** is by

$$\rho = \frac{V_4}{V_3}. \quad (2.33)$$

Exercise 15. Deduce the counterpart of (2.19):

$$P_5 = \left(\frac{\rho}{\varepsilon}\right)^k P_4, \quad T_5 = \left(\frac{\rho}{\varepsilon}\right)^{k-1} T_4. \tag{2.34}$$

Solution. The counterpart of (2.18) reads

$$P_4 {V_4}^k = P_5 {V_5}^k, \quad T_4 {V_4}^k = T_5 {V_5}^k. \tag{2.35}$$

Now we note that

$$\frac{V_4}{V_5} = \frac{V_4}{V_3} \cdot \frac{V_3}{V_1} = \frac{\rho}{\varepsilon}$$

by (2.6), (2.33). Substituting these in (2.35) proves (2.34).

Exercise 16. Use Exercise 15 to derive another expressions for (2.34):

$$P_5 = \sigma \rho^k P_1, \quad T_5 = \sigma \rho^k T_1. \tag{2.36}$$

Solution. Since $P_4 = P_3$, substituting (2.32), we obtain $P_5 = \rho P_2$. Now the first equality in (2.19) gives the first equality in (2.36). Similarly, using (2.33), we deduce the second equality.

Exercise 17. Derive the following equalities:

$$T_3 = \sigma {\varepsilon}^{k-1} T_1, \quad T_4 = \sigma \rho {\varepsilon}^{k-1} T_1. \tag{2.37}$$

Solution. Since $T_3 = T_2 \left(\frac{P_3}{P_2}\right) = \sigma T_2$, we have the first equality by (2.19) and (2.32). Similarly, from $T_4 = T_3 \left(\frac{V_4}{V_3}\right) = \rho T_3$ and (2.33), we deduce the second.

2.6 Heat efficiency

First we consider the **theoretical mean effective pressure** p_{ms} for the Sabathé cycle, which is defined to be the ratio of the theoretical work done on the cycle to the stroke volume $V_h = V_1 - V_2$. Suppose the gas is in m kg. Then in (2.12),

$$Q_1 = m\{c_V (T_3 - T_2) + C_P (T_4 - T_3)\}, Q_2 = m\{C_V (T_5 - T_1)\},$$

we obtain

$$W = \frac{mc_V}{A}\{(T_3 - T_2) + k(T_4 - T_3) - (T_5 - T_1)\}. \tag{2.38}$$

Substituting (2.19), (2.36) and (2.37) in (2.38), we deduce that

$$W = \frac{mc_V}{A}\varepsilon^{k-1}\left(\sigma - 1 + k\sigma(\rho - 1) - \frac{\sigma\rho^k - 1}{\varepsilon^{k-1}}\right) T_1. \tag{2.39}$$

Now substituting (2.16) in (2.39) and transforming $\frac{c_V}{c_P - c_V}$ into $\frac{1}{k-1}$, (2.39) becomes

$$W = mRc_V\varepsilon^{k-1}\left(\sigma - 1 + k\sigma(\rho - 1) - \frac{\sigma\rho^k - 1}{\varepsilon^{k-1}}\right)\frac{1}{k-1}T_1.$$

Dividing this by $V_1 - V_2$ and rewriting $\frac{mRT_1}{V_1 - V_2} = \frac{P_1V_1}{V_1 - V_2}$ as $P_1\frac{\varepsilon}{\varepsilon - 1}$, we obtain

Theorem 2.1. *The theoretical mean effective pressure of the Sabathé cycle is given by*

$$P_{ms} = \frac{\varepsilon^k\left(\sigma - 1 + k\sigma(\rho - 1)\right)}{(k-1)(\varepsilon - 1)}\left(1 - \frac{1}{\varepsilon^{k-1}}\frac{\sigma\rho^k - 1}{\sigma - 1 + k\sigma(\rho - 1)}\right)P_1. \quad (2.40)$$

Corollary 2.1. *The theoretical mean effective pressure of the Otto cycle is given by*

$$P_{mo} = \frac{\varepsilon^k(\sigma - 1)}{(k-1)(\varepsilon - 1)}\left(1 - \frac{1}{\varepsilon^{k-1}}\right)P_1, \quad (2.41)$$

while that of the Diesel cycle is given by

$$P_{md} = \frac{\varepsilon^k k(\rho - 1)}{(k-1)(\varepsilon - 1)}\left(1 - \frac{\rho^k - 1}{\varepsilon^{k-1}k(\rho - 1)}\right)P_1. \quad (2.42)$$

Chapter 3

Fluid dynamics: Navier-Stokes equation

Abstract

This chapter paves the foundations of fluid dynamics as a precursor of the study on the theory of lubrication, which is indispensable in reducing the friction—the most important challenge for human beings in the 21st century.

In §3.3 which is mostly based on [Kanemitsu *et al.* (2016)], [Li *et al.* (2018a)] we elucidate basic facts in fluid dynamics in [Khono (1989)], [Sena and Ohta (2007)], [Toyokura and Kamemoto (1976)] etc. We show that by the use of chain rule, Euler's form of velocity vectors is made clear. The main novelty of our method is a very effective use of differential forms coupled with the general form of Stokes' theorem to make many results in fluid mechanics simpler and clearer. In particular, we shall elucidate the notion of divergence and circulation in the 3-dimensional flow case in Theorem 3.6. Further in the case of 2-dimensional flow by the use of complex analysis, we reestablish the results in that theorem.

The main purpose of this chapter is to prove Theorem 3.12 to the effect that the Cauchy equation of motion implies the Navier-Stokes equation, thereby correcting the coefficients in [Grodins (1963)]. For this purpose we develop the vectorial version of the general Stokes theorem. From it we deduce the most important unicity theorem from which the equation of continuity follows immediately. In the same vein that the Cauchy equation of motion is a consequence of Newton's second law of motion follows immediately. We shall give a starter for the project of deriving the deepest results in vector analysis by the theory of generalized functions, which are known mainly as distributions. Our standpoint is, however, that of [Imai (1963)] which incorporates the Sato hyper-function as a main tool for interpreting the whirl flow.

For the sake of completeness we assemble some basic results from the theory of complex functions, cf. e.g. [Chakraborty *et al.* (2009)]. We state some basic well-known results from mathematical analysis for the sake of easy reference and self-containedness.

3.1 Introduction

§3.3 is based on [Kanemitsu *et al.* (2016)], where mathematical foundations of fluid dynamics were laid to some extent. Our main purpose is to prove Theorem 3.12 to the effect that the Cauchy equation of motion implies the

Navier-Stokes equation and also corrects the missing coefficients in [Grodins (1963)]. For this purpose we develop the vectorial version of the general Stokes theorem which itself is of interest in its own. From this we deduce the most important unicity theorem from which the equation of continuity immediately follows. In the same vein that the Cauchy equation of motion is a consequence of Newton's second law of motion follows immediately. The linear relation stated in the proof of Theorem 3.12 given in [Grodins (1963)] is rather misleading and we streamline the proof by a linear relation (3.99).

In the case of 2-dimensional flows, we revisit the Karman vortex street and reveal an unexpected hidden fact, via the partial fraction expansion of the cotangent function, that the expression (3.134) for the stream function for the Karman vortex street is, in the long run, a consequence of the functional equation for the Riemann zeta-function.

3.2 Fluid dynamics and lubrication

The flows are classified as in the following table.

Rheol. cons.	Dil. tens.	Temp. var.	Type
Gas	Incompressible	Steady	Viscous
Liquid	Compressible	Unsteady	Inviscid
State	div $= 0$	$\frac{\partial}{\partial t}\boldsymbol{v} = \boldsymbol{o}$	$\rho = 0$

Table 3.1: Classification of fluid flows

Motion char.	Spacial dim.
Irrotational	One dimensional
Rotational	Two dimensional
curl$\boldsymbol{v} = \boldsymbol{o}$	Three dimensional

Table 3.1: Classification of fluid flows (cont.)

3.3 Mathematical foundations of fluid dynamics

Let a vector-valued function $\boldsymbol{y} = \boldsymbol{y}(\boldsymbol{x})$ in a vector argument $\boldsymbol{x} = \begin{pmatrix} x_1 \\ x_2 \\ \vdots \\ x_n \end{pmatrix} \in$ $D \subset \mathbb{R}^n$ be given by

$$\begin{pmatrix} y_1 \\ y_2 \\ \vdots \\ y_m \end{pmatrix} = \boldsymbol{y} = \boldsymbol{y}(\boldsymbol{x}) = \begin{pmatrix} y_1(\boldsymbol{x}) \\ y_2(\boldsymbol{x}) \\ \vdots \\ y_m(\boldsymbol{x}) \end{pmatrix} = \begin{pmatrix} y_1(x_1, x_2, \cdots, x_n) \\ y_2(x_1, x_2, \cdots, x_n) \\ \cdots \\ y_m(x_1, x_2, \cdots, x_n) \end{pmatrix}, \tag{3.1}$$

which is equivalent to the system of equations

$$\begin{cases} y_1 = y_1(\boldsymbol{x}) = y_1(x_1, x_2, \cdots, x_n) \\ y_2 = y_2(\boldsymbol{x}) = y_2(x_1, x_2, \cdots, x_n) \\ \cdots \\ y_m = y_m(\boldsymbol{x}) = y_m(x_1, x_2, \cdots, x_n). \end{cases} \tag{3.2}$$

Definition 3.1. The vector-valued function $\boldsymbol{y} = \boldsymbol{y}(\boldsymbol{x})$ is said to be totally differentiable (or Fréchet differentiable) at $\boldsymbol{x}$ if

$$\boldsymbol{y}(\boldsymbol{x} + \boldsymbol{h}) = \boldsymbol{y}(\boldsymbol{x}) + A\boldsymbol{h} + o(|\boldsymbol{h}|) \tag{3.3}$$

as $\boldsymbol{h} \to \boldsymbol{o}$, i.e. $|\boldsymbol{h}| \to 0$. Here A is an (m, n)-matrix and is called the **gradient** (or the Jacobi matrix) of $\boldsymbol{y}$, denoted $\nabla \boldsymbol{y}$.

$$\nabla \boldsymbol{y} = \nabla \begin{pmatrix} y_1 \\ y_2 \\ \vdots \\ y_m \end{pmatrix} = \begin{pmatrix} \nabla y_1 \\ \nabla y_2 \\ \vdots \\ \nabla y_m \end{pmatrix} = \begin{pmatrix} \frac{\partial y_1}{\partial x_1} & \frac{\partial y_1}{\partial x_2} & \cdots & \frac{\partial y_1}{\partial x_n} \\ \frac{\partial y_2}{\partial x_1} & \frac{\partial y_2}{\partial x_2} & \cdots & \frac{\partial y_2}{\partial x_n} \\ \cdots \\ \frac{\partial y_m}{\partial x_1} & \frac{\partial y_m}{\partial x_2} & \cdots & \frac{\partial y_m}{\partial x_n} \end{pmatrix}. \tag{3.4}$$

In the case of a scalar function $\boldsymbol{y} = y = y(x_1, \cdots, x_n)$, the increment term $A\boldsymbol{h}$ in (3.1) amounts to

$$\mathrm{d}y = \mathrm{d}\boldsymbol{y} = \nabla y \mathrm{d}\boldsymbol{x} = \frac{\partial y}{\partial x_1}\mathrm{d}x_1 + \cdots + \frac{\partial y}{\partial x_n}\mathrm{d}x_n, \tag{3.5}$$

where $\mathrm{d}\boldsymbol{y}$ means the differential of $\boldsymbol{y}$, on writing $\boldsymbol{h} = \mathrm{d}\boldsymbol{x} = (\mathrm{d}x_1, \cdots, \mathrm{d}x_n)$.

$$\mathrm{d}\boldsymbol{y} = \mathrm{d}\boldsymbol{y}(\boldsymbol{x}) = \begin{pmatrix} \mathrm{d}y_1 \\ \mathrm{d}y_2 \\ \vdots \\ \mathrm{d}y_m \end{pmatrix} \tag{3.6}$$

(3.5) is a differential form of degree 1 in n variables, which will be used in the computation of differential of differential forms below.

Theorem 3.1. *Suppose* (3.1) *and*

$$\begin{pmatrix} z_1 \\ z_2 \\ \vdots \\ z_\ell \end{pmatrix} = \boldsymbol{z} = \boldsymbol{z}(\boldsymbol{y}) = \begin{pmatrix} z_1(\boldsymbol{y}) \\ z_2(\boldsymbol{y}) \\ \vdots \\ z_\ell(\boldsymbol{y}) \end{pmatrix} = \begin{pmatrix} z_1(y_1, y_2, \cdots, y_m) \\ z_2(y_1, y_2, \cdots, y_m) \\ \cdots \\ z_\ell(y_1, y_2, \cdots, y_m) \end{pmatrix}, \tag{3.7}$$

are differentiable. Then the composite function

$$(\boldsymbol{z} \circ \boldsymbol{y})(\boldsymbol{x}) = \boldsymbol{z}(\boldsymbol{y}(\boldsymbol{x})) = \begin{pmatrix} z_1(y_1(x_1, \cdots, x_n), \cdots, y_m(x_1, \cdots, x_n)) \\ z_2(y_1(x_1, \cdots, x_n), \cdots, y_m(x_1, \cdots, x_n)) \\ \cdots \\ z_\ell(y_1(x_1, \cdots, x_n), \cdots, y_m(x_1, \cdots, x_n)) \end{pmatrix} \tag{3.8}$$

is differentiable and the gradient is given as the product of the gradients

$$\nabla(\boldsymbol{z} \circ \boldsymbol{y}) = \nabla \boldsymbol{z} \nabla \boldsymbol{y}, \tag{3.9}$$

i.e. the chain rule holds true which reads componentwisely,

$$\frac{\partial z_i}{\partial x_j} = \sum_{k=1}^{m} \frac{\partial z_i}{\partial y_k} \frac{\partial y_k}{\partial x_j}. \tag{3.10}$$

3.4 Vector analysis

Definition 3.2. For a vector-valued function $\boldsymbol{f} = \begin{pmatrix} P \\ Q \\ R \end{pmatrix} \in C^1(X)$ (i.e. $P, Q, R \in C^1(X)$), we define its **divergence** $\operatorname{div} \boldsymbol{f}$ and **curl** $\operatorname{curl} \boldsymbol{f}$ (also called rotation $\operatorname{rot} \boldsymbol{f}$) by

$$\operatorname{div} \boldsymbol{f} = \frac{\partial P}{\partial x} + \frac{\partial Q}{\partial y} + \frac{\partial R}{\partial z} = \nabla \cdot \boldsymbol{f} \tag{3.11}$$

and

$$\operatorname{curl} \boldsymbol{f} = \begin{pmatrix} \begin{vmatrix} \frac{\partial}{\partial y} & Q \\ \frac{\partial}{\partial z} & R \end{vmatrix} \\ -\begin{vmatrix} \frac{\partial}{\partial x} & P \\ \frac{\partial}{\partial z} & R \end{vmatrix} \\ \begin{vmatrix} \frac{\partial}{\partial x} & P \\ \frac{\partial}{\partial y} & Q \end{vmatrix} \end{pmatrix} = \begin{pmatrix} \frac{\partial R}{\partial y} - \frac{\partial Q}{\partial z} \\ -\left(\frac{\partial R}{\partial x} - \frac{\partial P}{\partial z}\right) \\ \frac{\partial Q}{\partial x} - \frac{\partial P}{\partial y} \end{pmatrix} \tag{3.12}$$

$$= \nabla \times \boldsymbol{f}.$$

The formal inner product $\Delta = \nabla \cdot \nabla = \frac{\partial^2}{\partial x^2} + \frac{\partial^2}{\partial y^2} + \frac{\partial^2}{\partial z^2}$ is called the **Laplacian** and the solution ψ of the **Laplace equation**

$$\Delta \psi = \nabla \cdot \nabla \psi = \frac{\partial^2 \psi}{\partial x^2} + \frac{\partial^2 \psi}{\partial y^2} + \frac{\partial^2 \psi}{\partial z^2} = 0 \tag{3.13}$$

is called a **harmonic function**.

Throughout in what follows, we mean by a manifold a domain which is smooth enough to satisfy all the requirements in the statement. Therefore, one can think of it as a bounded domain with smooth boundary.

Theorem 3.2. (General Stokes' theorem) *Let $\mathcal{M}$ be a $(k+1)$-dimensional manifold, $\partial\mathcal{M}$ its boundary and let ω be a differential form of degree k in n variables. Then the identity*

$$\underbrace{\int \cdots \int}_{k+1}{}_{\mathcal{M}} \, \mathrm{d}\omega = \underbrace{\int \cdots \int}_{k}{}_{\partial\mathcal{M}} \, \omega \tag{3.14}$$

holds true or in the form

$$\mathrm{d}\omega(\mathcal{M}) = \omega(\partial\mathcal{M}). \tag{3.15}$$

name	degree	number of var.
General Stokes' theorem	k	n
Green's theorem	1	2
Gauss divergence theorem	2	3
Stokes' theorem	1	3

Table 3.2: Special cases of general Stokes' theorem

In what follows we apply the following rule for computing the product of differential forms.

$$\mathrm{d}y\mathrm{d}x = -\mathrm{d}x\mathrm{d}y, \quad \mathrm{d}x\mathrm{d}x = \mathrm{d}y\mathrm{d}y = 0. \tag{3.16}$$

The first equality in (3.16) may be thought of as representing the area with a sign of an infinitesimal parallelogram and the second as the area of a degenerated parallelogram.

Theorem 3.3. (Green's theorem) *Let $\mathcal{M} = \mathcal{D} \subset \mathbb{R}^2$ be a domain, $\partial\mathcal{M} = \partial\mathcal{D} \subset \mathbb{R}^2$ its boundary and let ω be a C^1 class differential form of degree 1 in 2 variables:*

$$\omega = P\mathrm{d}x + Q\mathrm{d}y. \tag{3.17}$$

Then we have

$$\iint_{\mathcal{D}} \mathrm{d}\omega = \int_{\partial\mathcal{D}} \omega \tag{3.18}$$

or more concretely,

$$\iint_{\mathcal{D}} \left(\frac{\partial Q}{\partial x} - \frac{\partial P}{\partial y}\right) \mathrm{d}x\mathrm{d}y = \int_{\partial\mathcal{D}} P\mathrm{d}x + Q\mathrm{d}y. \tag{3.19}$$

Proof. Since $\mathrm{d}\omega = \mathrm{d}P\mathrm{d}x + \mathrm{d}Q\mathrm{d}y$, the proof follows from (3.5) and the rule (3.16).

$$\mathrm{d}\omega = \mathrm{d}P\mathrm{d}x + \mathrm{d}Q\mathrm{d}y = \left(\frac{\partial P}{\partial x}\mathrm{d}x + \frac{\partial P}{\partial y}\mathrm{d}y\right)\mathrm{d}x + \left(\frac{\partial Q}{\partial x}\mathrm{d}x + \frac{\partial Q}{\partial y}\mathrm{d}y\right)\mathrm{d}y \tag{3.20}$$

$$= \frac{\partial P}{\partial y}\mathrm{d}y\mathrm{d}x + \frac{\partial Q}{\partial x}\mathrm{d}x\mathrm{d}y = \left(\frac{\partial Q}{\partial x} - \frac{\partial P}{\partial y}\right)\mathrm{d}x\mathrm{d}y.$$

□

Theorem 3.4. (Gauss divergence theorem) *Let $\mathcal{M} = R \subset \mathbb{R}^3$ be a domain, $\partial\mathcal{M} = \partial R \subset \mathbb{R}^3$ its boundary and let ω be a C^1 class differential form of degree 2 in 3 variables:*

$$\omega = P\mathrm{d}y\mathrm{d}z + Q\mathrm{d}z\mathrm{d}x + R\mathrm{d}x\mathrm{d}y. \tag{3.21}$$

Then we have

$$\iiint_{\mathcal{D}} \mathrm{d}\omega = \iint_{\partial\mathcal{D}} \omega \tag{3.22}$$

or more concretely,

$$\iiint_{\mathcal{D}} \left(\frac{\partial P}{\partial x} + \frac{\partial Q}{\partial y} + \frac{\partial R}{\partial z}\right) \mathrm{d}x\mathrm{d}y\mathrm{d}z = \iint_{\partial\mathcal{D}} P\mathrm{d}y\mathrm{d}z + Q\mathrm{d}z\mathrm{d}x + R\mathrm{d}x\mathrm{d}y. \tag{3.23}$$

Proof.

$$
\begin{aligned}
\mathrm{d}\omega &= \mathrm{d}P\mathrm{d}y\mathrm{d}x + \mathrm{d}Q\mathrm{d}z\mathrm{d}y + R\mathrm{d}x\mathrm{d}y \qquad (3.24)\\
&= \left(\frac{\partial P}{\partial x}\mathrm{d}x + \frac{\partial P}{\partial y}\mathrm{d}y + \frac{\partial P}{\partial y}\mathrm{d}z\right)\mathrm{d}y\mathrm{d}z + \left(\frac{\partial Q}{\partial x}\mathrm{d}x + \frac{\partial Q}{\partial y}\mathrm{d}y + \frac{\partial Q}{\partial y}\mathrm{d}z\right)\mathrm{d}z\mathrm{d}x\\
&+ \left(\frac{\partial R}{\partial x}\mathrm{d}x + \frac{\partial R}{\partial y}\mathrm{d}y + \frac{\partial R}{\partial y}\mathrm{d}z\right)\mathrm{d}x\mathrm{d}y\\
&= \frac{\partial P}{\partial x}\mathrm{d}x\mathrm{d}y\mathrm{d}z + \frac{\partial Q}{\partial y}\mathrm{d}y\mathrm{d}z\mathrm{d}x + \frac{\partial R}{\partial z}\mathrm{d}z\mathrm{d}x\mathrm{d}y.
\end{aligned}
$$

□

The name comes from the fact that the left-hand side of (3.23) is equal to the integral of divergence defined by (3.11).

Theorem 3.5. (Stokes' theorem) *Let $\mathcal{M} = \mathcal{D} \subset \mathbb{R}^3$ be a domain, $\partial\mathcal{M} = \partial\mathcal{D} \subset \mathbb{R}^3$ its boundary and let ω be a C^1 class differential form of degree* 1 *in* 3 *variables:*

$$\omega = P\mathrm{d}x + Q\mathrm{d}y + R\mathrm{d}z. \qquad (3.25)$$

Then we have

$$\iint_{\mathcal{D}} \mathrm{d}\omega = \int_{\partial\mathcal{D}} \omega \qquad (3.26)$$

or more concretely,

$$
\begin{aligned}
&\iint_{\mathcal{D}} \left(\frac{\partial Q}{\partial x} - \frac{\partial P}{\partial y}\right)\mathrm{d}x\mathrm{d}y + \iint_{\mathcal{D}} \left(\frac{\partial R}{\partial y} - \frac{\partial Q}{\partial z}\right)\mathrm{d}y\mathrm{d}z \qquad (3.27)\\
&+ \iint_{\mathcal{D}} \left(\frac{\partial P}{\partial z} - \frac{\partial R}{\partial x}\right)\mathrm{d}z\mathrm{d}x = \int_{\partial\mathcal{D}} P\mathrm{d}x + Q\mathrm{d}y + R\mathrm{d}z.
\end{aligned}
$$

3.5 Divergence

Let $\boldsymbol{v}$ be the vector field of velocity of the fluid flowing in a domain $\Omega \subset \mathbb{R}^3$ and let $X \subset \Omega$ be a bounded closed domain with its boundary ∂X forming a surface. Let ρ denote the density (distribution) function of the fluid in Ω. Then $\boldsymbol{f} = \rho\boldsymbol{v}$ is the vector field describing the flow of mass distribution of the fluid.

At a point $X \ni P : \boldsymbol{x} = \boldsymbol{x}(u, v)$, the normal vector $\boldsymbol{n}$ is given by

$$\boldsymbol{n} = \frac{1}{|\boldsymbol{x}_u \times \boldsymbol{x}_v|}\boldsymbol{x}_u \times \boldsymbol{x}_v \qquad (3.28)$$

and we write

$$f_{\boldsymbol{n}} = \boldsymbol{f} \cdot \boldsymbol{n} \tag{3.29}$$

for the component of $\boldsymbol{f}$ in the direction of $\boldsymbol{n}$. Hence $f_{\boldsymbol{n}}$ *expresses the ratio of mass flowing out at* P.

Therefore its integral over all ∂X

$$Q = Q(\partial X) := \iint_{\partial X} f_{\boldsymbol{n}} \tag{3.30}$$

expresses the **totality of masses flowing out of** X **in a unit time interval**.

Lemma 3.1. *Let* $\boldsymbol{f} = \begin{pmatrix} P \\ Q \\ R \end{pmatrix} \in C(\Omega)$ *be a continuous vector field in* $\Omega \subset \mathbb{R}^3$, $\omega = P\mathrm{d}y\mathrm{d}z + Q\mathrm{d}z\mathrm{d}x + R\mathrm{d}x\mathrm{d}y$ *a differential form of degree 2,* $S \subset \Omega$ *a smooth surface given by* $S : \boldsymbol{x} = \boldsymbol{x}(u, v)$, $(u, v) \in \mathbb{R}^2$. *Then we have*

$$\iint_S (P\mathrm{d}y\mathrm{d}z + Q\mathrm{d}z\mathrm{d}x + R\mathrm{d}x\mathrm{d}y) = \iint_S f_{\boldsymbol{n}}\,\mathrm{d}A, \tag{3.31}$$

where the right-hand side is the surface integral defined by

$$\iint_S f_{\boldsymbol{n}}\,\mathrm{d}A = \iint_D f_{\boldsymbol{n}}(\boldsymbol{x}(u, v))|\boldsymbol{x}_u \times \boldsymbol{x}_v|\,\mathrm{d}u\mathrm{d}v. \tag{3.32}$$

Proof. By (3.28) and (3.29), we have

$$f_{\boldsymbol{n}}|\boldsymbol{x}_u \times \boldsymbol{x}_v| = \boldsymbol{f} \cdot (\boldsymbol{x}_u \times \boldsymbol{x}_v). \tag{3.33}$$

Substituting this, we have

$$\iint_S f_{\boldsymbol{n}}\,\mathrm{d}A = \iint_D \left(P\frac{\partial(y, z)}{\partial(u, v)} - Q\frac{\partial(z, x)}{\partial(u, v)} + R\frac{\partial(x, y)}{\partial(u, v)} \right) \mathrm{d}u\mathrm{d}v, \tag{3.34}$$

which amounts to (3.31). □

Suppose there is no source or sink in X. Then since $Q(\partial X)$ in (3.30) indicates the rate of decrease of all masses in X, we have

$$\iint_{\partial X} f_{\boldsymbol{n}} = -\frac{\mathrm{d}}{\mathrm{d}t} \iiint_X \rho. \tag{3.35}$$

By Lemma 3.1 and Theorem 3.4, the left-hand side of (3.35) is $\iiint_X \operatorname{div} \boldsymbol{f}$:

$$\iint_{\partial X} f_{\boldsymbol{n}} = \iiint_X \operatorname{div} \boldsymbol{f}. \tag{3.36}$$

Hence

$$\iiint_X \operatorname{div} \boldsymbol{f} = -\frac{\mathrm{d}}{\mathrm{d}t} \iiint_X \rho. \tag{3.37}$$

If $\frac{\mathrm{d}\rho}{\mathrm{d}t}$ is continuous, then we may change the integration and differentiation on the right of (3.35) and we obtain

$$\iiint_X \operatorname{div} \boldsymbol{f} = - \iiint_X \frac{\partial \rho}{\partial t}. \tag{3.38}$$

Since (3.38) holds for any bounded domain X, we must have

$$\operatorname{div} \boldsymbol{f} = -\frac{\partial \rho}{\partial t}, \tag{3.39}$$

which is called the **equation of continuity** of fluid. This will be deduced again in Example 3.1 by the unicity theorem.

Definition 3.3. A fluid with constant density $\frac{\partial \rho}{\partial t} = 0$ is called an **incompressible fluid**. For an incompressible fluid, we have

$$\operatorname{div} \boldsymbol{f} = 0. \tag{3.40}$$

3.6 Circulation

Let C be a smooth Jordan curve and let $f_{\boldsymbol{t}}$ be the component of $\boldsymbol{f}$ in the direction of the tangent vector $\boldsymbol{t} = \dot{\boldsymbol{x}}(t)$ at a point $P \in C$:

$$f_{\boldsymbol{t}} = \boldsymbol{f} \cdot \frac{\boldsymbol{t}}{|\boldsymbol{t}|}. \tag{3.41}$$

Define the **circulation** of $\boldsymbol{f}$ around C by

$$\Gamma = \Gamma(C) = \int_C = \int_C f_{\boldsymbol{t}}\, \mathrm{d}s. \tag{3.42}$$

Lemma 3.2. *Let* $\boldsymbol{f} = \begin{pmatrix} P \\ Q \\ R \end{pmatrix} \in C(\Omega)$ *be a continuous vector field in* $\Omega \subset \mathbb{R}^3$, $\omega = P\mathrm{d}x + Q\mathrm{d}y + R\mathrm{d}z$ *a differential form of degree* 1, $C \subset \Omega$ *a smooth curve given by* $C : \boldsymbol{x} = \boldsymbol{x}(s)$, $s \in [0, \Lambda]$. *Then we have*

$$\int_C \boldsymbol{f} \cdot \mathrm{d}\boldsymbol{x} = \int_C P\mathrm{d}x + Q\mathrm{d}y + R\mathrm{d}z = \int_C f_{\boldsymbol{t}}\, \mathrm{d}s, \tag{3.43}$$

where the right-hand side is the 3*-dimensional line integral*

$$\int_C \boldsymbol{f} \cdot \mathrm{d}\boldsymbol{x} = \int_C \omega \tag{3.44}$$

and the right-hand side is an integral in the arc length.

Proof. We find that the vector

$$\boldsymbol{v} = \boldsymbol{v}(s) = \boldsymbol{x}'(s)$$

satisfies $|\boldsymbol{v}| = 1$. Hence the component in (3.41) is $f_{\boldsymbol{t}} = \boldsymbol{f} \cdot \boldsymbol{v}$. Hence

$$\int_C f_{\boldsymbol{t}}\, \mathrm{d}s = \int_C \left(P\frac{\mathrm{d}x}{\mathrm{d}s} + Q\frac{\mathrm{d}x}{\mathrm{d}s} + R\frac{\mathrm{d}x}{\mathrm{d}s} \right) \mathrm{d}s \tag{3.45}$$

which amounts to (3.44). □

Combined with Stokes' theorem, Theorem 3.5, this gives

$$\int_C f_{\boldsymbol{t}} = \iint_D \operatorname{curl} \boldsymbol{f}. \tag{3.46}$$

From (3.36) and (3.46) we deduce the following

Theorem 3.6. *If $X \subset \mathbb{R}^3$ is a bounded closed domain with its boundary ∂X forming a surface, then*

$$Q = Q(\partial X) = \iint_{\partial X} f_{\boldsymbol{n}} = \iiint_X \operatorname{div} \boldsymbol{f}. \tag{3.47}$$

If $D \subset \mathbb{R}^2$ is a domain with its boundary curve $\partial D = C$, then

$$\Gamma = \Gamma(C) = \int_C f_{\boldsymbol{t}} = \iint_D \operatorname{curl} \boldsymbol{f}. \tag{3.48}$$

Rheol. cons.	Dil. tens.	Temp. var.	Type
Gas	Incompressible	Steady	Viscous
Liquid	Compressible	Unsteady	Inviscid
State	div $= 0$	$\frac{\partial}{\partial t}\boldsymbol{v} = \boldsymbol{o}$	$\rho = 0$

Table 3.3: Classification of fluid flows

Motion char.	Spacial dim.
Irrotational	One dimensional
Rotational	Two dimensional
curl$\boldsymbol{v} = \boldsymbol{o}$	Three dimensional

Table 3.3: Classification of fluid flows (cont.)

3.7 2-dimensional flow

In this section we shall illustrate Theorem 3.6 by elucidating the results [Imai (1963), pp. 5-7] by complex analysis. Consider a 2-dimensional flow with its 3-dimensional velocity vector $\boldsymbol{v} = \begin{pmatrix} P \\ Q \\ 0 \end{pmatrix} \in C(D)$, where $D \subset \mathbb{R}^3$ is a domain with its boundary as a closed Jordan curve C. Then the divergence and the curl are defined respectively by

$$\operatorname{div} \boldsymbol{v} = \frac{\partial P}{\partial x} + \frac{\partial Q}{\partial y} \tag{3.49}$$

and

$$\operatorname{curl} \boldsymbol{v} = \begin{pmatrix} 0 \\ 0 \\ \omega \end{pmatrix}, \tag{3.50}$$

where $\omega = \frac{\partial Q}{\partial x} - \frac{\partial P}{\partial y}$ indicates the **vorticity**. By abuse of notation we understand $\operatorname{curl} \boldsymbol{v} = \omega$.

The 2-dimensional flow occurs when the fluid flows through a long cylindrical tube whose axis is perpendicular to the direction of the flow.

We consider the **complex velocity** corresponding to $\boldsymbol{v}$ in the form of the complex conjugate

$$\bar{v} = P - iQ. \tag{3.51}$$

This is because we want to have (3.54) valid.

We consider the function

$$W := \operatorname{curl} \boldsymbol{v} + i \operatorname{div} \boldsymbol{v} = \frac{\partial Q}{\partial x} - \frac{\partial P}{\partial y} + i \left(\frac{\partial P}{\partial x} + \frac{\partial Q}{\partial y} \right), \tag{3.52}$$

for which another expression is given in (3.67) later. We integrate it over D (with $\mathrm{d}A$ denoting the area element):

$$\iint_D W \,\mathrm{d}A = \iint_D \left(\frac{\partial Q}{\partial x} - \frac{\partial P}{\partial y} \right) \mathrm{d}A + i \iint_D \left(\frac{\partial P}{\partial x} + \frac{\partial Q}{\partial y} \right) \mathrm{d}A \tag{3.53}$$
$$= \int_C P\mathrm{d}x + Q\mathrm{d}y + i \int_C -Q\mathrm{d}x + P\mathrm{d}y$$

by Green's theorem.

We interpret the last two curvilinear integrals of (3.53) as a contour integral in the variable $z = x + iy$, whence $\mathrm{d}z = \mathrm{d}x + i\mathrm{d}y$. Hence

$$\bar{v}\,\mathrm{d}z = (P - iQ)(\mathrm{d}x + i\mathrm{d}y) = P\mathrm{d}x + Q\mathrm{d}y + i(-Q\mathrm{d}x + P\mathrm{d}y). \tag{3.54}$$

I.e.

$$\iint_D W \,\mathrm{d}A = \int_C \bar{v} \,\mathrm{d}z. \tag{3.55}$$

Since

$$v_t = P\frac{\mathrm{d}x}{\mathrm{d}s} + Q\frac{\mathrm{d}y}{\mathrm{d}s}, \quad v_n = -Q\frac{\mathrm{d}x}{\mathrm{d}s} + P\frac{\mathrm{d}y}{\mathrm{d}s}, \tag{3.56}$$

where $v_t = \boldsymbol{v} \cdot \boldsymbol{t}$ is the component in the direction of the tangent vector and v_n the component in the direction of the normal vector as defined in §3.5, we also have

$$\bar{v}\mathrm{d}z = (v_t + iv_n)\mathrm{d}s. \tag{3.57}$$

From (3.52), (3.55) and (3.57) we conclude that

$$\iint_D (\operatorname{curl} \boldsymbol{v} + i \operatorname{div} \boldsymbol{v}) \,\mathrm{d}A = \int_C (v_t + iv_n) \,\mathrm{d}s. \tag{3.58}$$

Comparing the real and imaginary parts, we conclude the 2-dimensional version of Theorem 3.6:

Theorem 3.7. *If $D \subset \mathbb{R}^2$ is a domain with its boundary curve $\partial D = C$, then*

$$Q = Q(\partial X) = \int_{\partial X} f_n = \iint_X \operatorname{div} \boldsymbol{f}. \tag{3.59}$$

$$\Gamma = \Gamma(C) = \int_C f_t = \iint_D \operatorname{curl} \boldsymbol{f}. \tag{3.60}$$

We now assume that the flow is incompressible and irrotational, i.e. div $\boldsymbol{v} = 0$ and curl $\boldsymbol{v} = 0$. Then by Theorem 3.15, we have $\int_C \bar{v} \,\mathrm{d}z = 0$ for any piecewise smooth Jordan curve C in D. Hence by Morera's theorem 3.19, $\bar{v}(z)$ is analytic in D. Then by Theorem 3.17, there exists a primitive function $\bar{v}(z)$, which is called the **complex potential** of $\bar{v}$.

A real-valued function ϕ such that $\boldsymbol{f} = \nabla \phi$ is called a **potential function** of $\boldsymbol{f}$. Thus the preceding argument gives rise to a potential function for the velocity $\bar{\boldsymbol{v}} = \begin{pmatrix} P \\ -Q \end{pmatrix}$ corresponding to $\bar{v}$ in (3.51).

3.8 Cauchy-Riemann equation

Theorem 3.8. *A function $w = f(z) = u + iv$ is analytic in the domain $D \subset \mathbb{C} \Longleftrightarrow u, v$ are totally differentiable on $D \subset \mathbb{R}^2$ and satisfy the* **Cauchy-Riemann equation**

$$\frac{\partial u}{\partial x} = \frac{\partial v}{\partial y}, \frac{\partial u}{\partial y} = -\frac{\partial v}{\partial x}; \qquad u_x = v_y, \, u_y = -v_x. \tag{3.61}$$

In this case, we have

$$f'(z) = \frac{\partial f}{\partial x} = u_x + iv_x. \tag{3.62}$$

(Also, substituting the Cauchy-Riemann equation in this formula, we may express it as $f'(z) = \frac{\partial f}{\partial (iy)} = -i(u_y + iv_y) = v_y - iu_y$.)

It follows that

$$\frac{\partial}{\partial z} f := \left(\frac{\partial}{\partial x} + i \frac{\partial}{\partial y} \right) f = \frac{\partial f}{\partial x} - \frac{\partial f}{\partial (iy)} = u_x - v_y + i(u_y + v_x). \tag{3.63}$$

Symbolically we may write

$$\frac{\partial}{\partial (x+iy)} = \frac{\partial}{\partial z}, \quad \frac{\partial}{\partial (x-iy)} = \frac{\partial}{\partial \bar{z}}. \tag{3.64}$$

Hence

Theorem 3.9. *Suppose $P, Q \in C^2(D)$, where $D \subset \mathbb{R}^2$ is a domain. Then $f(z) = P + iQ$ is analytic in $D \subset \mathbb{C}$ if and only if*

$$\frac{\partial}{\partial z} f = 0. \tag{3.65}$$

This follows from the fact that (3.65) is equivalent to the Cauchy-Riemann equation.

Hence

$$i \frac{\partial}{\partial \bar{z}} \bar{v} = i \left(\frac{\partial}{\partial x} - i \frac{\partial}{\partial y} \right) (P - iQ) = \frac{\partial Q}{\partial x} - \frac{\partial P}{\partial y} + i \left(\frac{\partial P}{\partial x} + \frac{\partial Q}{\partial y} \right). \tag{3.66}$$

Comparing this with (3.52), we conclude that

$$W = \operatorname{curl} \boldsymbol{v} + i \operatorname{div} \boldsymbol{v} = i \frac{\partial}{\partial \bar{z}} \bar{v}. \tag{3.67}$$

It follows from Theorem 3.9 that if $\frac{\partial}{\partial \bar{z}} \bar{v}$ is analytic in a domain D with a closed curve C as its boundary, then

$$\int_C (\operatorname{curl} \boldsymbol{v} + i \operatorname{div} \boldsymbol{v}) \, \mathrm{d}z = i \int_C \frac{\partial}{\partial \bar{z}} \bar{v} \, \mathrm{d}z = 0, \tag{3.68}$$

whence that the fluid is incompressible and irrotational.

However, if it has a pole in D, as in the case of the **vortex filament** $i\frac{k}{z}$, where $k > 0$ is a constant, then the integral along the curve C encircling the pole is not zero but the residue at its pole as furnished by Theorem 3.20. In the case of the vortex filament, the residue at the origin is $-2\pi k$ and (3.68) reads

$$\int_C \operatorname{curl} \boldsymbol{v}\, \mathrm{d}z + i \int_C \operatorname{div} \boldsymbol{v}\, \mathrm{d}z = -2\pi k, \tag{3.69}$$

whence $\int_C \operatorname{curl} \boldsymbol{v}\, \mathrm{d}z = -2\pi k \neq 0$, i.e. there is a whirl at the pole. We are led to the study of such flows by the theory of Sato hyperfunctions, cf. [Imai (1963)] and §3.14.

3.9 Extension of Reynolds theory

Banerjee, Shandil and Katyal [Banerjee *et al.* (1986)] unified the theory for two different forms of a flow given in the following table.

Regime	flow	forces	Reynolds theory appl.
upper	turbulent	-	×
narrow intermediate	laminar	viscous, inertia	○
lower	laminar	viscous	○

Table 3.4: Types of flows according to the size of the Reynolds number

In order to understand the setting, one needs to know the governing equations for steady flows. However, the equations are given in an abbreviated form, it is desirable to state the full form of the Navier-Stokes equation and in this section we give some rudiments of that equation.

3.10 Vectorial version of Stokes' theorem

Definition 3.4. We extend the definition of the divergence to a matrix. Let

$$P = (\boldsymbol{p}_1, \cdots, \boldsymbol{p}_n), \tag{3.70}$$

where $\boldsymbol{p}_j, 1 \le j \le n$ are of C^1 class as in Definition 3.2. Then

$$\nabla \cdot P = \nabla \cdot (\boldsymbol{p}_1, \cdots, \boldsymbol{p}_n) = (\nabla \cdot \boldsymbol{p}_1, \cdots, \nabla \cdot \boldsymbol{p}_n). \tag{3.71}$$

Theorem 3.10. (General Stokes' theorem) *Let $\mathcal{M}$ be a $(k+1)$-dimensional manifold, $\partial\mathcal{M}$ its boundary and let ω be a differential form of degree k in n variables. Then the identity*

$$\underbrace{\int\cdots\int}_{k+1}{}_{\mathcal{M}}\, \mathrm{d}\omega = \underbrace{\int\cdots\int}_{k}{}_{\partial\mathcal{M}}\, \omega \tag{3.72}$$

holds true; in more concise form it reads

$$\mathrm{d}\omega(\mathcal{M}) = \omega(\partial\mathcal{M}). \tag{3.73}$$

Furthermore, suppose $\omega = \boldsymbol{f}\cdot \mathrm{d}A$ with 'area element' $A = A(\partial\mathcal{M})$, we have $\mathrm{d}\omega = \nabla\cdot\boldsymbol{f}\mathrm{d}V$ with 'volume element' $V = V(\mathcal{M})$. Then (3.72) *reads*

$$\underbrace{\int\cdots\int}_{k+1}{}_{\mathcal{M}}\, \nabla\cdot\boldsymbol{f}\mathrm{d}V = \underbrace{\int\cdots\int}_{k}{}_{\partial\mathcal{M}}\, \boldsymbol{f}\cdot\mathrm{d}A. \tag{3.74}$$

In vectorial form, (3.74) *reads*

$$\underbrace{\int\cdots\int}_{k+1}{}_{\mathcal{M}}\, \nabla\cdot\boldsymbol{P}\mathrm{d}V = \underbrace{\int\cdots\int}_{k}{}_{\partial\mathcal{M}}\, \boldsymbol{P}\cdot\mathrm{d}A, \tag{3.75}$$

where $\boldsymbol{P} = (\boldsymbol{p}_1, \cdots, \boldsymbol{p}_n)$ is an $n\times n$-matrix, under the inner product rule

$$\boldsymbol{P}\cdot\mathrm{d}A = (\boldsymbol{p}_1\cdot\mathrm{d}A, \cdots, \boldsymbol{p}_n\cdot\mathrm{d}A). \tag{3.76}$$

Corollary 3.1. (Unicity theorem) *Suppose $n = k+1$ and that we are given the identity*

$$\underbrace{\int\cdots\int}_{k+1}{}_{\mathcal{M}}\, \nabla\cdot\boldsymbol{P}\mathrm{d}V = \underbrace{\int\cdots\int}_{k+1}{}_{\mathcal{M}}\, \nabla\cdot\boldsymbol{Q}\mathrm{d}V \tag{3.77}$$

for any bounded domain $V\subset\mathcal{M}$. Then

$$\boldsymbol{P} = \boldsymbol{Q} \tag{3.78}$$

throughout $\mathcal{M}$. Furthermore, the condition (3.77) *may be replaced by*

$$\underbrace{\int\cdots\int}_{k+1}{}_{\mathcal{M}}\, \nabla\cdot\boldsymbol{P}\mathrm{d}V = \underbrace{\int\cdots\int}_{k}{}_{\partial\mathcal{M}}\, \boldsymbol{Q}\cdot\mathrm{d}A. \tag{3.79}$$

Proof. Equation (3.77) amounts to

$$\underbrace{\int \cdots \int}_{k+1} {}_{\mathcal{M}}\, \nabla \cdot (\boldsymbol{P} - \boldsymbol{Q}) \mathrm{d}V = 0 \tag{3.80}$$

for any bounded domain V. Since the integrand is continuous, if it is non-zero at a point $\boldsymbol{z}_0$, say $= c > 0$. Then in a neighborhood V_0 of the point it is positive and so the integral over V_0 is positive, a contradiction.

By (3.75), the second term on the right of (3.79) is $\underbrace{\int \cdots \int}_{k+1} {}_{\mathcal{M}}\, \nabla \cdot \boldsymbol{Q} \mathrm{d}V$ leading to (3.77).

Given (3.79), the general Stokes theorem implies (3.77) and we still have the validity of the corollary, completing the proof. □

Remark 3.1. The condition $n = k + 1$ indicates the saturation of the integrals in (3.77), i.e. they are scalar integrals including the case (3.75) of matrices. In view of (3.71) the latter case amounts to

$$\left(\underbrace{\int \cdots \int}_{k+1} {}_{\mathcal{M}} \nabla \cdot \boldsymbol{p}_1, \cdots, \underbrace{\int \cdots \int}_{k+1} {}_{\mathcal{M}} \nabla \cdot \boldsymbol{p}_n \right)$$

$$= \left(\underbrace{\int \cdots \int}_{k+1} {}_{\mathcal{M}} \nabla \cdot \boldsymbol{q}_1, \cdots, \underbrace{\int \cdots \int}_{k+1} {}_{\mathcal{M}} \nabla \cdot \boldsymbol{q}_n \right),$$

which implies coincidence of components, and so the unicity still holds.

Example 3.1. (Equation of continuity) We reprove (3.39) from unicity theorem, Corollary 3.1. Let $\boldsymbol{v}$ be the vector field of velocity of the fluid flowing in a domain $\Omega \subset \mathbb{R}^3$ and let $X \subset \Omega$ be a bounded closed domain with its boundary ∂X forming a surface. Let ρ denote the continuous density (distribution) function of the fluid in Ω. Then $\boldsymbol{f} = \rho \boldsymbol{v}$ is the vector field describing the flow of mass distribution of the fluid. Suppose there is no source or sink in X. Then in the same way as we proved (3.39), we arrive at

$$\iiint_X \operatorname{div} \boldsymbol{f} = \iiint_X \left(-\frac{\partial \rho}{\partial t} \right). \tag{3.81}$$

Hence, by Corollary 3.1

$$\operatorname{div} \boldsymbol{f} = -\frac{\partial \rho}{\partial t}, \tag{3.82}$$

the **equation of continuity** of fluid again.

Example 3.2. (Cauchy equation of motion) In this example we show that Newton's second law of motion entails the Cauchy equation of motion (3.87).

We consider a flow in a domain $\Omega \subset \mathbb{R}^3$ and let $V \subset \Omega$ be a bounded closed domain with its boundary surface $A = \partial V$. Let ρ denote the continuous density (distribution) function of the fluid in Ω. Then Newton's second law of motion (1.6) reads

$$M\boldsymbol{a} = \boldsymbol{F}, \tag{3.83}$$

where $M = \int_V \rho \, \mathrm{d}V$ indicates the mass and $\boldsymbol{F}$ is the sum of all external forces acting on the fluid volume. The force consists of body force $\boldsymbol{F}_b$ and surface force $\boldsymbol{F}_s$. The body forces act on the center of mass of the fluid volume and we assume it to be the gravitational force

$$\boldsymbol{F}_b = \int_V \rho \boldsymbol{g} \, \mathrm{d}V \tag{3.84}$$

and the surface force is given by stress dyadic

$$\boldsymbol{F}_s = \int_A \boldsymbol{P} \cdot \mathrm{d}A = \int_V \nabla \cdot \boldsymbol{P} \, \mathrm{d}V. \tag{3.85}$$

Hence

$$\int_V (\boldsymbol{g} - \boldsymbol{a})\rho \mathrm{d}V + \int_V \nabla \cdot \boldsymbol{P} \, \mathrm{d}V = 0. \tag{3.86}$$

Corollary 3.1 implies the **Cauchy equation of motion**

$$(\boldsymbol{g} - \boldsymbol{a})\rho + \nabla \cdot P = \boldsymbol{o}. \tag{3.87}$$

We state well-known special cases given in Table 3.5 of the general Stokes theorem in [Kanemitsu *et al.* (2016)].

name	degree	number of var.
General Stokes' theorem	k	n
Green's theorem	1	2
Gauss divergence theorem	2	3
Stokes' theorem	1	3

Table 3.5: Special cases of the general Stokes' theorem

3.11 Navier-Stokes equation

However, in fluid dynamics it is customary to follow Euler to assume that the velocity components are functions of the position components as well as that of the time.

By the chain rule,

$$\begin{aligned}\frac{\mathrm{d}u}{\mathrm{d}t} &= \frac{\partial u}{\partial t} + \frac{\partial u}{\partial x}\frac{\mathrm{d}x}{\mathrm{d}t} + \frac{\partial u}{\partial y}\frac{\mathrm{d}y}{\mathrm{d}t} + \frac{\partial u}{\partial z}\frac{\mathrm{d}z}{\mathrm{d}t} \\ &= \frac{\partial u}{\partial t} + \frac{\partial u}{\partial x}u + \frac{\partial u}{\partial y}v + \frac{\partial u}{\partial z}w, \\ \frac{\mathrm{d}v}{\mathrm{d}t} &= \frac{\partial v}{\partial t} + \frac{\partial v}{\partial x}u + \frac{\partial v}{\partial y}v + \frac{\partial v}{\partial z}w, \\ \frac{\mathrm{d}w}{\mathrm{d}t} &= \frac{\partial w}{\partial t} + \frac{\partial w}{\partial x}u + \frac{\partial w}{\partial y}v + \frac{\partial w}{\partial z}w.\end{aligned} \tag{3.88}$$

Therefore, the acceleration $\boldsymbol{a} = \boldsymbol{a}(x, y, z, t)$ is to be understood as

$$\boldsymbol{a} = \frac{\partial}{\partial t}\boldsymbol{v} + \boldsymbol{v} \cdot \nabla \boldsymbol{v}. \tag{3.89}$$

We confine ourselves to the isotropical fluids, i.e. those which behave in the same way for all directions.

Definition 3.5. Let P denote the **stress dyadic** given by

$$P = \begin{pmatrix} p_{xx} & p_{xy} & p_{xz} \\ p_{yx} & p_{yy} & p_{yz} \\ p_{zx} & p_{zy} & p_{zz} \end{pmatrix}, \tag{3.90}$$

where in the notation p_{ij}, the subscripts i, j take on values x, y, z and the outer subscript indicates the direction of stress and the inner one indicates the plane on which stress acts.

Let $\dot{S}$ denote the **strain rate dyadic** given by

$$\begin{aligned}\dot{S} = \frac{\partial}{\partial t}\epsilon_{ij} &= \begin{pmatrix} \dot{\epsilon}_{xx} & \dot{\epsilon}_{xy} & \dot{\epsilon}_{xz} \\ \dot{\epsilon}_{yx} & \dot{\epsilon}_{yy} & \dot{\epsilon}_{yz} \\ \dot{\epsilon}_{zx} & \dot{\epsilon}_{zy} & \dot{\epsilon}_{zz} \end{pmatrix} \\ &= \begin{pmatrix} \frac{\partial u}{\partial x} & \frac{1}{2}\left(\frac{\partial u}{\partial y} + \frac{\partial v}{\partial x}\right) & \frac{1}{2}\left(\frac{\partial u}{\partial z} + \frac{\partial w}{\partial x}\right) \\ \frac{1}{2}\left(\frac{\partial v}{\partial x} + \frac{\partial u}{\partial y}\right) & \frac{\partial v}{\partial y} & \frac{1}{2}\left(\frac{\partial v}{\partial z} + \frac{\partial w}{\partial y}\right) \\ \frac{1}{2}\left(\frac{\partial w}{\partial x} + \frac{\partial u}{\partial z}\right) & \frac{1}{2}\left(\frac{\partial w}{\partial y} + \frac{\partial v}{\partial z}\right) & \frac{\partial w}{\partial z} \end{pmatrix}.\end{aligned} \tag{3.91}$$

Note that $\dot{S}$ is a symmetric matrix.

Then Stokes proved

Theorem 3.11.

$$P = 2\mu\dot{S} + bE, \quad b = -\left(p + \frac{2}{3}\mu\,\mathrm{div}(\boldsymbol{v})\right), \tag{3.92}$$

where μ indicates the **viscosity** *of the fluid and E is the identity matrix of degree 3. Or in terms of components*

$$\begin{aligned} p_{xx} &= -p - \frac{2}{3}\mu\left(\frac{\partial u}{\partial x} + \frac{\partial v}{\partial y} + \frac{\partial w}{\partial z}\right) + 2\mu\frac{\partial u}{\partial x} \\ p_{yy} &= -p - \frac{2}{3}\mu\left(\frac{\partial u}{\partial x} + \frac{\partial v}{\partial y} + \frac{\partial w}{\partial z}\right) + 2\mu\frac{\partial v}{\partial y} \\ p_{zz} &= -p - \frac{2}{3}\mu\left(\frac{\partial u}{\partial x} + \frac{\partial v}{\partial y} + \frac{\partial w}{\partial z}\right) + 2\mu\frac{\partial w}{\partial z} \end{aligned} \tag{3.93}$$

$$\begin{aligned} p_{yz} &= p_{zy} = \mu\left(\frac{\partial w}{\partial y} + \frac{\partial v}{\partial z}\right) \\ p_{zx} &= p_{xz} = \mu\left(\frac{\partial u}{\partial z} + \frac{\partial w}{\partial x}\right) \\ p_{xy} &= p_{yx} = \mu\left(\frac{\partial v}{\partial x} + \frac{\partial u}{\partial y}\right). \end{aligned} \tag{3.94}$$

Proof. We write

$$P_1 = \begin{pmatrix} p_{xx} \\ p_{xy} \\ p_{xz} \\ p_{yy} \\ p_{yz} \\ p_{zz} \end{pmatrix}, \tag{3.95}$$

$$\dot{S}_1 = \begin{pmatrix} \dot{\epsilon}_{xx} \\ \dot{\epsilon}_{xy} \\ \dot{\epsilon}_{xz} \\ \dot{\epsilon}_{yy} \\ \dot{\epsilon}_{yz} \\ \dot{\epsilon}_{zz} \end{pmatrix}, \tag{3.96}$$

$$\boldsymbol{b} = \begin{pmatrix} b_1 \\ 0 \\ b_2 \\ 0 \\ b_3 \\ 0 \end{pmatrix}, \tag{3.97}$$

and

$$A = (a_i j)_{1 \leq i,j \leq 6} \tag{3.98}$$

and consider the linear relation

$$P_1 = A\dot{S}_1 + \boldsymbol{b}. \tag{3.99}$$

Stokes proved that

$$a_{11} = a_{44} = a_{66} := a, \tag{3.100}$$

say,

$$a_{22} = a_{33} = a_{55} = \frac{1}{2}(a_{11} - a_{14}), \tag{3.101}$$

$$b_1 = b_2 = b_3 := b, \tag{3.102}$$

say, and

$$\frac{1}{2}(a_{11} - a_{14}) = \mu \tag{3.103}$$

and that all other constants are zero. It follows that A is diagonal.

Recall that the pressure p at a point in a fluid is, by definition, the negative of the arithmetic mean of the three normal stresses p_{ij} acting on three mutually perpendicular surface elements:

$$p = -\frac{1}{3}(p_{xx} + p_{yy} + p_{zz}). \tag{3.104}$$

Now comparing the traces of P and $\dot{S}$, we obtain

$$p_{xx} + p_{yy} + p_{zz} = a(\dot{\epsilon}_{xx} + \dot{\epsilon}_{yy} + \dot{\epsilon}_{zz}) + 3b = a\nabla \cdot \boldsymbol{v} + 3b, \tag{3.105}$$

in which Stokes found that

$$a = 2\mu. \tag{3.106}$$

Since the left-hand side is $-3p$, it follows that

$$b = -p - \frac{1}{3}\mu\nabla \cdot \boldsymbol{v} = -(p + \frac{2}{3}\mu\nabla \cdot \boldsymbol{v}). \tag{3.107}$$

Hence

$$\boldsymbol{b} = -(p + \frac{2}{3}a\nabla \cdot \boldsymbol{v})E. \tag{3.108}$$

Incorporating (3.103) and (3.106), we find that the diagonal of A is

$$A = \begin{pmatrix} 2\mu \\ \mu \\ \mu \\ 2\mu \\ \mu \\ 2\mu \end{pmatrix}. \tag{3.109}$$

This completes the proof. □

The **Navier-Stokes equation** for general flow reads

$$\frac{\partial}{\partial t}\boldsymbol{v} + \boldsymbol{v}\cdot\nabla\boldsymbol{v} = \boldsymbol{g} - \frac{1}{\rho}\nabla p + \nu\Delta\boldsymbol{v} + \frac{1}{3}\nu\frac{\partial}{\partial\boldsymbol{x}}\operatorname{div}(\boldsymbol{v}), \tag{3.110}$$

where $\boldsymbol{g} = \begin{pmatrix} X \\ Y \\ Z \end{pmatrix}$ is the mass force, p is the pressure given by (3.104) above, $\nu = \frac{\mu}{\rho}$ is the **kinematic viscosity**, which is the ratio of the viscosity to the density of the fluid, and Δ is the Laplacian defined by (3.13).

In component form, (3.110) reads

$$\rho\left(\frac{\partial u}{\partial t} + u\frac{\partial u}{\partial x} + v\frac{\partial u}{\partial y} + w\frac{\partial u}{\partial z}\right) = \rho X - \frac{\partial p}{\partial x} + \mu\left(\Delta\, u + \frac{1}{3}\frac{\partial\operatorname{div}(\boldsymbol{v})}{\partial x}\right) \tag{3.111}$$

$$\rho\left(\frac{\partial v}{\partial t} + u\frac{\partial v}{\partial x} + v\frac{\partial v}{\partial y} + w\frac{\partial v}{\partial z}\right) = \rho Y - \frac{\partial p}{\partial y} + \mu\left(\Delta\, v + \frac{1}{3}\frac{\partial\operatorname{div}(\boldsymbol{v})}{\partial y}\right)$$

$$\rho\left(\frac{\partial w}{\partial t} + u\frac{\partial w}{\partial x} + v\frac{\partial w}{\partial y} + w\frac{\partial w}{\partial z}\right) = \rho Z - \frac{\partial p}{\partial z} + \mu\left(\Delta\, w + \frac{1}{3}\frac{\partial\operatorname{div}(\boldsymbol{v})}{\partial z}\right).$$

Remark 3.2. Note that (3.111) corrects [Granger (1985), p. 198] in which there is a missing coefficient $\frac{1}{3}$. Also the statement on [Granger (1985), p. 181] to the effect that one can consider the most general form of a linear relation between a stress and rate of strain as

$$\mathbf{P} = a\dot{\mathbf{S}} + bE \tag{3.112}$$

is rather misleading since what is considered is the system of linear equations on [Granger (1985), p. 182], which is nothing other than (3.99).

Theorem 3.12. *The Cauchy equation of motion entails the Navier-Stokes equation.*

Proof. In view of the Cauchy equation of motion (3.87), to prove the theorem it suffices to compute the components of $\nabla \cdot P$.

$$
\begin{aligned}
&\frac{\partial}{\partial x}p_{xx} + \frac{\partial}{\partial y}p_{yx} + \frac{\partial}{\partial z}p_{zx} \qquad (3.113)\\
&= \frac{\partial}{\partial x}\left(-p + \frac{4}{3}\mu\frac{\partial u}{\partial x} - \frac{2}{3}\mu\left(\frac{\partial v}{\partial y} + \frac{\partial w}{\partial z}\right)\right)\\
&+ \mu\frac{\partial}{\partial y}\left(\frac{\partial v}{\partial x} + \frac{\partial u}{\partial y}\right) + \mu\frac{\partial}{\partial z}\left(\frac{\partial w}{\partial x} + \frac{\partial u}{\partial z}\right)\\
&= -\frac{\partial p}{\partial x} + \mu\left(\frac{4}{3}\frac{\partial^2 u}{\partial x^2} - \frac{2}{3}\frac{\partial^2 v}{\partial x\partial y} - \frac{2}{3}\frac{\partial^2 w}{\partial x\partial z} + \frac{\partial^2 v}{\partial y\partial x} + \frac{\partial^2 u}{\partial y^2} + \frac{\partial^2 u}{\partial z^2} + \frac{\partial^2 w}{\partial x\partial z}\right)\\
&= -\frac{\partial p}{\partial x} + \mu\left(\Delta u + \frac{1}{3}\left(\frac{\partial^2 u}{\partial x^2} + \frac{\partial^2 v}{\partial x\partial y} + \frac{\partial^2 w}{\partial x\partial z}\right)\right)\\
&= -\frac{\partial p}{\partial x} + \mu\left(\Delta u + \frac{1}{3}\frac{\partial}{\partial x}\left(\frac{\partial u}{\partial x} + \frac{\partial v}{\partial y} + \frac{\partial w}{\partial z}\right)\right)\\
&= -\frac{\partial p}{\partial x} + \mu\left(\Delta u + \frac{1}{3}\frac{\partial}{\partial x}\nabla\cdot\boldsymbol{v}\right).
\end{aligned}
$$

$$
\begin{aligned}
&\frac{\partial}{\partial x}p_{xy} + \frac{\partial}{\partial y}p_{yy} + \frac{\partial}{\partial z}p_{zy} \qquad (3.114)\\
&= \mu\frac{\partial}{\partial x}\left(\frac{\partial v}{\partial x} + \frac{\partial u}{\partial y}\right)\\
&+ \frac{\partial}{\partial y}\left(-p + \frac{4}{3}\mu\frac{\partial v}{\partial y} - \frac{2}{3}\mu\left(\frac{\partial u}{\partial x} + \frac{\partial w}{\partial z}\right)\right) + \mu\frac{\partial}{\partial z}\left(\frac{\partial w}{\partial y} + \frac{\partial v}{\partial z}\right)\\
&= -\frac{\partial p}{\partial y} + \mu\left(\frac{4}{3}\frac{\partial^2 v}{\partial y^2} - \frac{2}{3}\frac{\partial^2 u}{\partial x\partial y} - \frac{2}{3}\frac{\partial^2 w}{\partial y\partial z} + \frac{\partial^2 v}{\partial x^2} + \frac{\partial^2 u}{\partial x\partial y} + \frac{\partial^2 w}{\partial y\partial z} + \frac{\partial^2 v}{\partial z^2}\right)\\
&= -\frac{\partial p}{\partial y} + \mu\left(\Delta v + \frac{1}{3}\left(\frac{\partial^2 u}{\partial x\partial y} + \frac{\partial^2 v}{\partial y^2} + \frac{\partial^2 w}{\partial y\partial z}\right)\right)\\
&= -\frac{\partial p}{\partial y} + \mu\left(\Delta v + \frac{1}{3}\frac{\partial}{\partial y}\left(\frac{\partial u}{\partial x} + \frac{\partial v}{\partial y} + \frac{\partial w}{\partial z}\right)\right)\\
&= -\frac{\partial p}{\partial y} + \mu\left(\Delta v + \frac{1}{3}\frac{\partial}{\partial y}\nabla\cdot\boldsymbol{v}\right).
\end{aligned}
$$

$$
\begin{aligned}
&\frac{\partial}{\partial x}p_{xz} + \frac{\partial}{\partial y}p_{yz} + \frac{\partial}{\partial z}p_{zz} \qquad (3.115)\\
&= \mu\frac{\partial}{\partial x}\left(\frac{\partial u}{\partial z} + \frac{\partial w}{\partial x}\right)\\
&+ \mu\frac{\partial}{\partial y}\left(\frac{\partial v}{\partial z} + \frac{\partial w}{\partial y}\right) + \frac{\partial}{\partial z}\left(-p + \frac{4}{3}\mu\frac{\partial w}{\partial z} - \frac{2}{3}\mu\left(\frac{\partial u}{\partial x} + \frac{\partial v}{\partial y}\right)\right)\\
&= -\frac{\partial p}{\partial z} + \mu\left(\frac{4}{3}\frac{\partial^2 w}{\partial z^2} - \frac{2}{3}\frac{\partial^2 u}{\partial z \partial x} - \frac{2}{3}\frac{\partial^2 v}{\partial y \partial z} + \frac{\partial^2 u}{\partial z \partial x} + \frac{\partial^2 w}{\partial x^2} + \frac{\partial^2 v}{\partial y \partial z} + \frac{\partial^2 w}{\partial y^2}\right)\\
&= -\frac{\partial p}{\partial z} + \mu\left(\Delta w + \frac{1}{3}\left(\frac{\partial w}{\partial^2 z} + \frac{\partial^2 u}{\partial z \partial x} + \frac{\partial^2 v}{\partial y \partial z}\right)\right)\\
&= -\frac{\partial p}{\partial z} + \mu\left(\Delta w + \frac{1}{3}\frac{\partial}{\partial z}\left(\frac{\partial u}{\partial x} + \frac{\partial v}{\partial y} + \frac{\partial w}{\partial z}\right)\right)\\
&= -\frac{\partial p}{\partial z} + \mu\left(\Delta w + \frac{1}{3}\frac{\partial}{\partial z}\nabla \cdot \boldsymbol{v}\right).
\end{aligned}
$$

Substituting these equalities in (3.87) proves (3.111), thereby proving the theorem. □

In case the mass force is negligible, the Navier-Stokes equation for an incompressible flow reads

$$
\frac{\partial}{\partial t}\boldsymbol{v} + \boldsymbol{v} \cdot \nabla \boldsymbol{v} = -\frac{1}{\rho}\nabla p + \nu \Delta \boldsymbol{v}. \qquad (3.116)
$$

Theorem 3.13. *Newton's second law of motion entails the Cauchy equation of motion, which in turn entails the Navier-Stokes equation.*

This follows from Theorem 3.12 and Example 3.2.

3.12 Appearance of the Reynolds number

According to the size of the Reynolds number, there appear different stages of a steady flow, cf. e.g. [Banerjee *et al.* (1986)].

Regime	flow	forces	Reynolds theory appl.
upper	turbulent	-	×
narrow intermediate	laminar	viscous, inertia	∘
lower	laminar	viscous	∘

Table 3.6: Types of flows according to the size of the Reynolds number

We are mainly concerned with the Karman vortex street introduced below and the contents are partly modified and essentially strengthened form of [Chakraborty *et al.* (2009), Example 26, §7]. Recall the definition of the complex logarithm function

$$\log w = z = x + iy = \log|w| + i \arg w = \log|w| + i \arctan\frac{v}{u}, \tag{3.117}$$

where the last expression is valid for $u \neq 0$.

Example 3.3. (2-dimensional flow of a vortex) Consider the complex potential given by $w = -\frac{\Gamma}{2\pi i}\log z$, where $\Gamma > 0$ is a constant. Putting $w = u + iv$ and $z = re^{i\vartheta}$, we have by (3.117)

$$\varphi = -\frac{\Gamma}{2\pi}\vartheta, \quad \psi = \frac{\Gamma}{2\pi}\log r. \tag{3.118}$$

The stream line $\psi =$ const. is $r =$ const., i.e. the concentric circles around the origin. This is the flow (in the positive direction) around the vortex at the origin as we see presently. The radiation velocity $v(r)$, the tangential velocity $v(\vartheta)$ are given respectively by

$$v(r) = -\frac{\partial\varphi}{\partial r}, \quad v(\vartheta) = -\frac{1}{r}\frac{\partial\varphi}{\partial\vartheta}.$$

Hence, in our case, by (3.118)

$$v(r) = 0, \quad v(\vartheta) = \frac{\Gamma}{2\pi r}, \tag{3.119}$$

so that the **induced velocity** caused by the vortex is only the circulation $v(\vartheta)$ around the origin. The constant Γ turns out to be the circulation and is called the **intensity** of the flow.

Apparently, if the vortex is at the point α, the complex potential is given by

$$w = -\frac{\Gamma}{2\pi i}\log(z - \alpha). \tag{3.120}$$

3.13 The Karman vortex row

First consider the case of a **vortex row** in which there are infinitely many vortices with the same intensity and rotating in the positive direction at distance a apart. We consider the $2n+1$ vortices with the central one at the origin. Then by (3.120) and the principle of superposition, the complex potential is

$$w_n = -\frac{\Gamma}{2\pi i}\left(\log z + \sum_{k=1}^{n}(\log(z - ka) + \log(z + ka))\right). \tag{3.121}$$

Lemma 3.3. *The limit of* (3.121)

$$w = \lim_{n\to\infty} w_n = -\frac{\Gamma}{2\pi i}\left(\log z + \sum{}' \log(z - an)e^{\frac{z}{an}}\right) \tag{3.122}$$

can be expressed as

$$w = -\frac{\Gamma}{2\pi i}\log\sin\frac{\pi z}{a}. \tag{3.123}$$

Proof. We invoke the partial fraction expansion of the cotangent function valid for all $z \in \mathbb{C}$ save for multiples of π:

$$\begin{aligned} \cot z &= \frac{1}{z} + \sum_{n=-\infty}^{\infty}{}' \left(\frac{1}{z-\pi n} + \frac{1}{\pi n}\right) \\ &= \frac{1}{z} + 2z\sum_{n=1}^{\infty}\frac{1}{z^2 - \pi^2 n^2}. \end{aligned} \tag{3.124}$$

Since the series is uniformly convergent in any domain not containing multiples of π, we may integrate the first equality of (3.124) term by term from 1 to z to obtain

$$\log\sin z = \log z + \sum{}' \log(z - \pi n)e^{\frac{z}{\pi n}}, \tag{3.125}$$

which we may interpret to mean

$$\sin z = z\prod_{n=-\infty}^{\infty}{}' \left(1 - \frac{z}{\pi n}\right)e^{\frac{z}{\pi n}}. \tag{3.126}$$

□

Definition 3.6. The Hurwitz zeta-function is defined as the perturbed Dirichlet series

$$\zeta(s, a) = \sum_{n=1}^{\infty}\frac{1}{(n+a)^s}, \tag{3.127}$$

where $\operatorname{Re} s > 1$ and a is a constant $\operatorname{Re} a > 0$. Its counterpart is the polylogarithm function

$$l_s(a) = \sum_{n=1}^{\infty}\frac{e^{2\pi i n a}}{n^s}. \tag{3.128}$$

The special case $a = 1$ of the Hurwitz zeta-function or the polylogarithm function is the celebrated Riemann zeta-function

$$\zeta(s) = \sum_{n=1}^{\infty}\frac{1}{n^s}. \tag{3.129}$$

These functions can be continued meromorphically over the whole plane via their symmetry—functional equations.

Lemma 3.4. [Kanemitsu and Tsukada (2007), Theorem 5.5] *The following formulas are all equivalent.* (i) *The asymmetric form of the functional equation for the Riemann zeta-function:*

$$\zeta(1-s) = 2^{1-s}\pi^{-s}\,\Gamma(s)\cos\Big(\frac{\pi s}{2}\Big)\,\zeta(s). \tag{3.130}$$

(ii) *The functional equation for the Hurwitz zeta-function (or the Hurwitz formula)* $(0 < a < 1)$

$$\zeta(s,a) = -i\,(2\pi)^{s-1}\,\Gamma(1-s)\big(e^{\frac{\pi}{2}is}\,l_{1-s}(a) - e^{-\frac{\pi}{2}is}\,l_{1-s}(1-a)\big). \tag{3.131}$$

(iii) *The partial fraction expansion of the cotangent function* (3.124) *stated above.*

(iv) *The product representation for the sine function*

$$\frac{\sin \pi z}{\pi z} = \prod_{n=1}^{\infty}\left(1 - \frac{z^2}{n^2}\right), \tag{3.132}$$

which is a proper form of (3.126).

With (3.123) in mind, we may consider the **Karman vortex street** which is the row of two rows of infinite vortices with the same intensity but with the opposite rotation. We express them as two rows above and below the real axis at the height h and each vortex in the one row lies in the middle of the other row. The complex potential of the above vortex row is given by

$$w = -\frac{\Gamma}{2\pi i}\log\sin\frac{\pi}{a}\left(z - \frac{ih}{2}\right),$$

and that of the lower row is

$$w = -\frac{\Gamma}{2\pi i}\log\sin\frac{\pi}{a}\left(z + \frac{ih}{2} + \frac{a}{2}\right).$$

Theorem 3.14. *The expression for the complex potential of the Karman vortex street*

$$w = -\frac{\Gamma}{2\pi i}\log\frac{\sin\frac{\pi}{a}\left(z - \frac{ih}{2}\right)}{\cos\frac{\pi}{a}\left(z + \frac{ih}{2}\right)} \tag{3.133}$$

is, in a long run, is a consequence of the functional equation for the Riemann zeta-function, and a fortiori, such is the expression for the stream function ψ, *being the imaginary part of* (3.133),

$$w = \frac{\Gamma}{4\pi}\log\frac{\cosh\frac{2\pi}{a}\left(y - \frac{h}{2}\right) - \cos\frac{2\pi}{a}x}{\cosh\frac{2\pi}{a}\left(y + \frac{h}{2}\right) + \cos\frac{2\pi}{a}x}. \tag{3.134}$$

Proof follows immediately from Lemmas 3.3 and 3.4.

It is known due to Karman that the vortex street is stable when $\sin \frac{\pi h}{a} = 1$ or $\frac{h}{a} \approx 0.281$.

Let ν be the kinematic coefficient of viscosity of the fluid defined above. Let r be the radius of the cylinder and let U be the uniform speed of the flow. Then the **Reynolds number** R_e is defined by

$$R_e = \frac{2rU}{\nu}. \tag{3.135}$$

It was shown by Roshko (1953) that the Karman vortex street arises for $R_e \geq 40$.

Remark 3.3. In most of the textbooks on fluid mechanics, one finds the following argument: Since $\log(z - ka) + \log(z + ka) = \log(z^2 - ka^2)$, it follows, after slight transformation, that

$$w_n = -\frac{\Gamma}{2\pi i}\left(\log \frac{\pi z}{a} \prod_{k=1}^{n}\left(1 - \frac{z^2}{k^2 a^2}\right)\right) + \text{const.}$$

Hence choosing the coordinates which annihilate the constant and letting $n \to \infty$, we obtain

$$w = \lim_{n\to\infty} w_n = -\frac{\Gamma}{2\pi i}\left(\log \frac{\pi z}{a} \prod_{k=1}^{\infty}\left(1 - \frac{z^2}{k^2 a^2}\right)\right). \tag{3.136}$$

Comparing (3.136) with (3.126), we conclude (3.123).
Of course, this is fallacious and we deduce (3.123) using the second equality of (3.124).

Proof of (3.134). In view of the form of (3.133), the imaginary part can be computed from $\log |\frac{\sin z}{\cos z'}|$ and it suffices to find $|\sin(x + iy)|$. Since

$$\sin(x + iy) = \frac{1}{2i}\left(\left(e^x - e^{-x}\right)\cos y + \left(e^x + e^{-x}\right)\sin y\right),$$

it follows that

$$|\sin(x + iy)|^2 = 2\left(e^x - e^{-x}\right) + 2\left(\sin^2 y - \cos^2 y\right) = 4(\cosh 2x - \cos 2y).$$

Similarly,

$$|\cos(x + iy)|^2 = 4(\cosh 2x + \cos 2y).$$

Equation (3.134) immediately follows from these. □

3.14 Description of a whirl by the Sato hyperfunction

There is a distribution-theoretic treatment of fluid mechanics for which we refer to [Imai (1963)]. For distributions cf. [Donoghue (1969)], [Lighthill (1958)], [Papoulis (1962)], and [Yoshida (1956)]. In Imai [Imai (1963)], there is a description of a whirl but the following is clearer. In order to present it, we recall basic facts about the projective limit.

Definition 3.7. Let M be a partially ordered set. Then M is called a **directed set** if given two elements μ_1, μ_2, there exists an element ν such that $\mu_1 \leq\neq, \mu_2 \leq \nu$. A commutative diagram $\{X_\mu, \rho_\nu^\mu\}$ with M its type which consists of sets X_μ and maps ρ_ν^μ is called an **inductive system** or a direct system. I.e., suppose for each $\mu \in$ M, there corresponds a set X_μ and for every pair (μ, ν) with $\mu \leq \nu$, there exists a map

$$\rho_\nu^\mu : X_\mu \to X_\nu \tag{3.137}$$

satisfying

$$\rho_\mu^\mu = I_{X_\mu}, \quad \rho_\nu^\mu \rho_\mu^\lambda = \rho_\nu^\lambda \quad (\lambda \leq \mu \leq \nu). \tag{3.138}$$

Then in the union $\cup X_\mu$ we introduce the equivalence relation: two elements $x_{\mu_1} \in X_{\mu_1}$ and $x_{\mu_2} \in X_{\mu_2}$ are equivalent

$$x_{\mu_1} \sim x_{\mu_2} \tag{3.139}$$

if for some $\nu \geq \mu_1, \mu_2$

$$\rho_\nu^{\mu_1} x_{\mu_1} = \rho_\nu^{\mu_2} x_{\mu_2} \tag{3.140}$$

holds true.

We call the equivalence class $\cup X_\mu / \sim$ the **inductive limit** (or direct limit).

We are now in a position to follow Hiramatsu and Saito [Hiramatsu and Saito (2016)] to give a definition of the Sato hyperfunction.

For any open set $D \subset \mathbb{C}$, let $H(D)$ denote the ring of all holomorphic functions on D. Let S be any locally closed subset of $\mathbb{R}$. We denote by $N(S)$ the family of all complex neighborhoods of S, i.e., all open sets of $\mathbb{C}$ containing S as a closed subset. Also we denote by $H(S)$ and $\tilde{H}(S)$ the inductive limits (as defined above) of $\{H(D) : D \in N(S)\}$ and $\{H(D-S) : D \in N(S)\}$ with respect to the canonical homomorphisms, respectively. $\tilde{H}(S)$ is regarded as an extension ring of $H(S)$ in a natural manner, and so an $H(S)$-module $M(S)$ may be defined by

$$M(S) \equiv \tilde{H}(S) \pmod{H(S)}.$$

Each element of $M(S)$ is called a **hyperfunction** on S. By definition, each hyperfunction $g \in M(S)$ is represented by an element of $\tilde{H}(S)$ and therefore by an element $\varphi \in H(D-S)$ with some $D \in N(S)$. We call such φ a *defining function* of g and write

$$g = [\varphi, D] = [\varphi]$$

or

$$g(x) = [\varphi(z), D]_{z=x} = [\varphi(z)]_{z=x}.$$

We set $I = D \cap \mathbb{R}$ and define $\varphi(x \pm i0) \in M(I)$, the boundary values of $\varphi(z) \in H(D-I)$, by

$$\varphi(x+i0) = [\varepsilon\varphi, D],$$
$$\varphi(x-i0) = -[\bar{\varepsilon}\varphi, D]$$

with

$$\varepsilon(z) = \begin{cases} 1 \ (\operatorname{Im} z > 0) \\ 0 \ (\operatorname{Im} z < 0), \end{cases}$$

$$\bar{\varepsilon}(z) = \varepsilon(-z) = \begin{cases} 0 \ (\operatorname{Im} z > 0) \\ 1 \ (\operatorname{Im} z < 0). \end{cases}$$

Then we have the following representation of $g \in [\varphi, D]$:

$$g(x) = \varphi(x+i0) - \varphi(x-i0).$$

This describes well the whirl at x. We return to a description of the vortex flow in §3.13 elsewhere.

3.15 Appendix C: Some facts from complex analysis

Theorem 3.15. (Cauchy integral theorem) *If $f(z)$ is one-valued and analytic in a simply-connected domain $D \subset \mathbb{C}$, then for any piecewise smooth closed Jordan curve $C \subset D$, we have*

$$\int_C f(z)\,\mathrm{d}z = 0. \tag{3.141}$$

Proof. Let $f = u + iv$. We apply Theorem 3.3 to

$$\int_C f(z)\,\mathrm{d}z = \int_C u\mathrm{d}x - v\mathrm{d}y + i\int_C v\mathrm{d}x + u\mathrm{d}y \tag{3.142}$$

to deduce ($X = \operatorname{Int} C \cup C$)

$$\int_C f(z)\,\mathrm{d}z = \int\int_X -\left(\frac{\partial v}{\partial x} + \frac{\partial u}{\partial y}\right)\mathrm{d}x\mathrm{d}y + i\int\int_X \left(\frac{\partial u}{\partial x} - \frac{\partial v}{\partial y}\right)\mathrm{d}x\mathrm{d}y, \tag{3.143}$$

in which both of the double integrals are 0 by the Cauchy-Riemann equation, completing the proof. □

To be able to apply Green's Theorem, we need to assume that u, v are of $C^1(X)$ class. There was tremendous endeavor made toward the generalization of the Cauchy integral theorem whose general form is the following.

Theorem 3.16. *Let C be any rectifiable Jordan closed curve with its finite interior D. If $f(z)$ is bounded and analytic on D and continuous on $\bar{D} = D \cup C$, then* (3.141) *holds true.*

The proof uses the method of approximating C by a polygon $C' \in D$ and proving that the integrals along C and C' can be made as close as we please in view of continuity of $f(z)$ on $\bar{D}$.

For practical use, however, integration paths are piecewise smooth ones combining polygons, arcs etc. and the functions considered are mostly smooth, we may take the applicability of the integral theorem for granted.

Theorem 3.15 is equivalent to the following two fundamental theorems. This indicates that the Cauchy integral theorem is of utmost importance in complex analysis but since primitive functions cannot in general be expressed in elementary functions, Theorem 3.18 is of little practical use for evaluating the value of an integral. It is residue calculus that shows its power in computing the values of integrals.

Theorem 3.17. (Existence of a primitive function) *If $f(z)$ is one-valued and analytic in a simply-connected domain $D \subset \mathbb{C}$, then there exists a function $F(z)$ analytic in D such that*

$$F'(z) = f(z). \tag{3.144}$$

A primitive function is unique up to a constant (integral constant).

As in calculus, the uniqueness is a consequence of the Newton-Leibniz rule.

Theorem 3.18. (Computation of definite integrals) *If $f(z)$ is one-valued and analytic in a simply-connected domain $D \subset \mathbb{C}$, then for any curve $C : z = z(t), t \in [a, b]$ in D of finite length, we have*

$$\int_C f(z)\,\mathrm{d}z = F(z(b)) - F(z(a)), \tag{3.145}$$

where $F(z)$ is a primitive function of $f(z)$.

The Cauchy integral theorem has a converse given by

Theorem 3.19. (Morera) *If $f(z)$ is continuous in a simply-connected domain $D \subset \mathbb{C}$ and*

$$\int_C f(z)\,\mathrm{d}z = 0 \tag{3.146}$$

for any piecewise smooth Jordan closed curve C in D, then $f(z)$ is analytic in D.

Theorem 3.20. (**Cauchy residue theorem**) *Let C be a piecewise smooth closed Jordan curve whose interior is D and let $f(z)$ be analytic in D except for a finite number of poles at α_k, $1 \le k \le n$. Then*

$$\int_C f(z)\mathrm{d}z = 2\pi i \sum_{k=1}^{n} \operatorname*{Res}_{z=\alpha_k} f(z).$$

The Cauchy integral theorem 3.15 may be thought of as a special case of the Cauchy residue theorem with no pole in the domain.

Fig. 3.1: Iluminati ot syuda

Embroidering words are to indicate the passage in cyrilic. Iluminati ot syuda

Iluminati = ilumination, ot = from, syuda = here
Altogether, "light from here."

Chapter 4

Control theory

Abstract

One of the primary concerns in [Li *et al.* (2015)] was the control of EV participation in GG, and in particular the study of linear systems and elucidation of various state equations in [Hossain *et al.* (2014)] etc. In all scientific disciplines, all roads lead to the **concatenation of two components**. In our case they are the impedances, which we view as **operators**. E.g. $Z_C i = \frac{1}{C}\int_0^t i\,dt$ in the case of a condenser with capacitance C. When considering a cascade connection of two impedances Z_1, Z_2, we understand the sum $Z_1 + Z_2$ of operators applies to the current i flowing through them as $Z_1 i + Z_2 i$. From this point of view, things become far easier and we may be able to clarify all the state equations in a very apparent fashion. We view all the systems as embedded in the state equation, thus visualizing the state. As has been demonstrated briefly in [Li *et al.* (2012), Chapter 6], for a unified treatment of cascade connections, the chain scattering representation of the plant a là [Kimura (1997)] is quite useful. Since the inverters are of three-phase type, there is a possibility of applying it. The chain scattering representation includes the feedback connection in a natural way and we may consider the H^∞-control problem in this framework. In the unit feedback system also included is the PID-compensator. Both of the H^∞- and PID-controllers are applied successfully in the EV control by Cao [Cao and Cao (2006)], [Ye *et al.* (2007)] etc., which we may unify in our scope.

The state is an interface between the past and the present and we may view all controls and possible cyberattack assessment as a **dynamical system**. Through communications system, the grid functions may be improved, just as modern cars, whether gasoline-propelled or not, are highly controlled by installed computers. In the case of power grid, the present grid has almost no fear of cyberattack. Controlled grid which is a must for the human being to survive, there will be increased vulnerabilities just as with the present Internet environment since there will be many components connected including many EVs through wireless communications systems. We hope to study these dynamical systems from the point of view of the *transitional state* [Rozenfel'd and Yahinson (1966)] in which the Mikusinski distributions are used as opposed to the Sato hyperfunction in §3.14. It is quite natural to incorporate the distributions or hyperfunctions wherever the delta-function or similar non-functions appear.

Studying smart grid operations has significant similarities to that of EVs in that both essentially depend on computer control, many monitoring meters are equipped so that the system should run in order, and there is a **self-healing** function being installed so that the new system can predict any possible danger and can prevent or at least mitigate it when it occurs, giving an option for the system operator (resp. driver) to manually handle the danger, when appropriate. In the future system, it is hoped that new analytical programs are used which can perceive any possible menace to the system and as soon as they perceive it, they take an automatic action to prevent it or in the worst case, if it occurs, liquidate the effect to its minimum.

Indeed, what we expect will be the advanced form of a smart gird, Green Grid, in which the main source of energy supply will be from the use of solar power including wind power (being caused by solar power). It is crucial to have advanced batteries which can be charged at rather low voltage, e.g. by the light of the setting sun for the Green Grid to come true.

4.1 State space representation and visualization principle

Let $\boldsymbol{x} = \boldsymbol{x}(t) \in \mathbb{R}^n$, $\boldsymbol{u} = \boldsymbol{u}(t) \in \mathbb{R}^r$ and $\boldsymbol{y} = \boldsymbol{y}(t) \in \mathbb{R}^m$ be the **state** function, **input** function and **output** function, respectively. We write $\dot{\boldsymbol{x}}$ for $\frac{\mathrm{d}}{\mathrm{d}t}\boldsymbol{x}$.

Then a **state equation** for a **linear system** is usually given as the system of DEs (differential equations)

$$\begin{cases} \dot{\boldsymbol{x}} &= A\boldsymbol{x} + B\boldsymbol{u}, \\ \boldsymbol{y} &= C\boldsymbol{x} + D\boldsymbol{u} \end{cases} \tag{4.1}$$

where $A \in M_{n,n}(\mathbb{R})$, B, C, D are given constant matrices.

Using the matrix exponential function e^{At}, the first equation in (1.249) can be solved to be of the same form as (1.249):

$$x = e^{At}x(0) + Be^{At}\int_0^t e^{-A\tau}u(t)\,\mathrm{d}\tau. \tag{4.2}$$

It is usually the case that if we work in the (complex) frequency domain, the things are much easier. The Laplace transform has the effect of shifting from time domain to frequency domain and as a version of the Fourier transform, it is invertible, i.e. it restores the information in frequency domain in time domain. Cf. §1.5.2 and §8.3 for some more material on Laplace transform. Taking the Laplace transform of (4.1) with $\boldsymbol{x}(0) = \boldsymbol{o}$, we obtain

$$\begin{cases} sX(s) = AX(s) + BU(s) \\ \ \ Y(s) = CX(s) + DU(s), \end{cases} \tag{4.3}$$

which we solve as

$$Y(s) = G(s)U(s), \tag{4.4}$$

where

$$G(s) = C(sI - A)^{-1}B + D, \tag{4.5}$$

where I indicates the identity matrix, which is sometimes denoted I_n to show its size.

In general, supposing that the initial values of all the signals in a system are 0, we call the ratio of output/input of the signal, the **transfer function**, and denote it by $G(s)$, $\Phi(s)$, etc. We may suppose so because if the system is in equilibrium, then we may take the values of parameters at that moment as standard and may suppose the initial values to be 0. In control theory one works with the transfer functions which are rational functions and gives back the information to time domain.

(4.5) is called the **state space representation** (form, realization, description, characterization) of the transfer function $G(s)$ of the system (4.1), and is written as

$$G(s) = \left(\begin{array}{c|c} A & B \\ \hline C & D \end{array}\right). \tag{4.6}$$

4.2 Accommodation of some control plants

The three main ingredients in (electrical) circuits are coil (L), condenser (C) and resistance (R). The inverse electro-motive force generated by these component is given respectively by

$$e_L = -L\frac{\mathrm{d}i}{\mathrm{d}t}, \quad e_C = -\int_0^t i\,\mathrm{d}t, \quad e_R = -Ri, \tag{4.7}$$

where $i = i(t)$ and $e = e(t)$ with subscript indicates the current and the voltage of the prescribed component resp. Hereafter we write u for the voltage e.

The governing law of the circuits is the **Kirchhoff laws** which have two versions. The first law is the one used for node analysis to the effect that the sum of currents flowing into a node is 0 while the second law is the one used for loop analysis which is to the effect that the sum of all electro-motive forces in a closed circuit is 0.

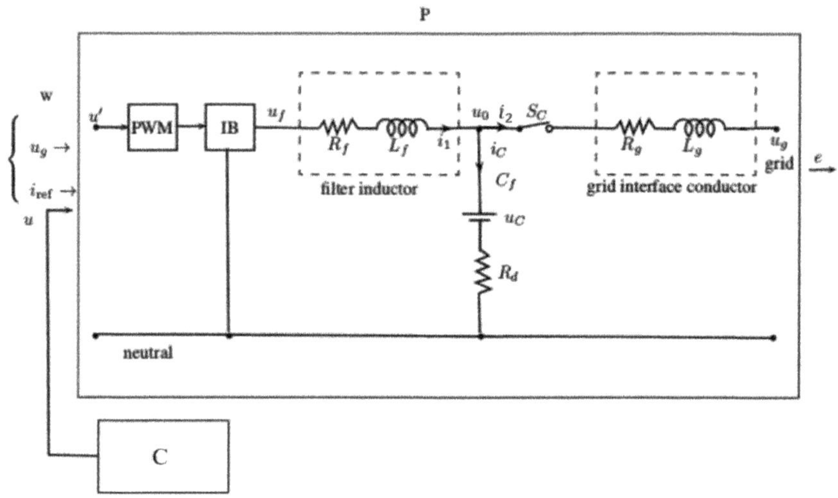

Fig. 4.1: Control plant P

Here PWM = Power Width Modulation and IB = Inverter Bridge. Modified from [Zong and Hornik (2014), pp. 83-84].

First consider the LC filter consisting of L_f, C_f in Fig. 4.1. As a consequence of the Kirchhoff law we have

$$i_1 = i_C + i_2. \tag{4.8}$$

The potential difference $u_0 - u_f$ and $-u_0 = 0 - u_0$ are given respectively by

$$u_0 - u_f = -R_f i_1 - L_f \frac{\mathrm{d}i_1}{\mathrm{d}t} \tag{4.9}$$

and

$$-u_0 = -u_C - R_d i_C, \tag{4.10}$$

where $u_C = \int_0^t i_C \,\mathrm{d}t$.

Hence we conclude that

$$R_f i_1 + L_f \frac{\mathrm{d}i_1}{\mathrm{d}t} = u_f - u_C - R_d i_C = u_f - u_C - R_d(i_1 - i_2), \tag{4.11}$$

whence that

$$\frac{\mathrm{d}i_1}{\mathrm{d}t} = \frac{1}{L_f}\left(-(R_f + R_d)i_1 + R_d i_2 - u_C + u_f\right) \tag{4.12}$$

or

$$\frac{\mathrm{d}i_1}{\mathrm{d}t} = \frac{1}{L_f}\left(-(R_f + R_d)i_1 + R_d i_2 - u_C + u_g + u\right) \tag{4.13}$$

on using the relation $u_f = u + u_g$.

Also for the grid filter consisting of L_g, C_f, the potential difference $u_g - u_0$ is given by

$$u_g - u_0 = -R_g i_2 - L_g \frac{\mathrm{d}i_2}{\mathrm{d}t}, \tag{4.14}$$

so that from (4.10), we obtain

$$R_g i_2 + L_g \frac{\mathrm{d}i_2}{\mathrm{d}t} = u_C + R_d(i_1 - i_2) - u_g, \tag{4.15}$$

whence

$$\frac{\mathrm{d}i_2}{\mathrm{d}t} = \frac{1}{L_g}\left(R_d i_1 - (R_g + R_d)i_2 + u_C - u_g\right). \tag{4.16}$$

Finally

$$\frac{\mathrm{d}u_C}{\mathrm{d}t} = i_C = i_1 - i_2. \tag{4.17}$$

We may express the control plant P in Fig. 4.1 in the form of (4.1). Let

$$\boldsymbol{x} = \begin{pmatrix} i_1 \\ i_2 \\ u_c \end{pmatrix}, \quad \boldsymbol{u} = (\boldsymbol{w}, u), \quad \boldsymbol{w} = \begin{pmatrix} u_g \\ i_{\text{ref}} \end{pmatrix}, \quad y = e = i_{\text{ref}} - i_2 \tag{4.18}$$

be the state variable, the external input and the output, respectively. Hence, writing

$$A = \begin{pmatrix} \frac{R_f+R_d}{L_f} & \frac{R_d}{L_f} & -\frac{1}{L_f} \\ \frac{R_d}{L_g} & -\frac{R_g+R_d}{L_f} & \frac{1}{L_g} \\ \frac{1}{C_f} & -\frac{1}{C_f} & 0 \end{pmatrix}, \quad B = (B_1, B_2), \quad B_1 = \begin{pmatrix} \frac{1}{L_f} & 0 \\ -\frac{1}{L_f} & 0 \\ 0 & 0 \end{pmatrix}, \tag{4.19}$$

$$B_2 = \begin{pmatrix} \frac{1}{L_f} \\ 0 \\ 0 \end{pmatrix}, \quad C_1 = (-R_d, R_d, 0), \quad D_1 = (0, 1), \quad D_2 = 0,$$

we may combine (4.13), (4.16) and (4.17) to give

$$\begin{cases} \dot{\boldsymbol{x}} &= A\boldsymbol{x} + B_1\boldsymbol{w} + B_2\boldsymbol{u}, \\ \boldsymbol{y} &= C_1\boldsymbol{x} + D_1\boldsymbol{w} + D_2\boldsymbol{u} \end{cases} \tag{4.20}$$

Viewing B, D as block decompositions, (4.20) amounts to (4.1).

Exercise 18. If the external input in (4.18) is $\boldsymbol{w} = \begin{pmatrix} u_g \\ u_{\text{ref}} \end{pmatrix}$, then we still have the state equation (4.20) with (4.19) changed by

$$A = \begin{pmatrix} -\frac{R_f+R_d}{L_f} & \frac{R_d}{L_f} & -\frac{1}{L_f} \\ \frac{R_d}{L_g} & -\frac{R_g+R_d}{L_f} & \frac{1}{L_g} \\ 0 & -\frac{1}{C_f} & 0 \end{pmatrix}, \quad B = (B_1, B_2), \quad B_1 = \begin{pmatrix} \frac{1}{L_f} & 0 \\ -\frac{1}{L_f} & 0 \\ 0 & 0 \end{pmatrix}, \tag{4.21}$$

$$B_2 = \begin{pmatrix} \frac{1}{L_f} \\ 0 \\ 0 \end{pmatrix}, \quad C_1 = (0, -1, 0), \quad D_1 = (0, 1), \quad D_2 = 0.$$

Solution. It is a challenge to keep both the THD (**Total Harmonic Distortion**) low of the **inverter local load voltage** u_0 and the **grid current** (the current i_2 flowing through the grid interface inductor). The inverter LCL plant (control plant P) may be thought of as a **cascaded control structure** consisting of an inner loop voltage controller and an outer loop of current controller, cf. Fig. 4.3.

Let

$$\boldsymbol{x}_v = \begin{pmatrix} i_1 \\ u_c \end{pmatrix}, \quad \boldsymbol{w}_v = (\boldsymbol{w}, u), \quad \boldsymbol{w}_v = \begin{pmatrix} i_2 \\ u_{\text{ref}} \end{pmatrix}, \quad y_v = e_v = u_{\text{ref}} - u_0 \tag{4.22}$$

be the state variable, the external input and the output signal (which is the tracking error), respectively.

The reasoning is similar to the one given for (4.20). Hence we conclude that

$$R_f i_1 + L_f \frac{\mathrm{d}i_1}{\mathrm{d}t} = u_f - u_C - R_d i_C = u_f - u_C - R_d(i_1 - i_2), \tag{4.23}$$

whence that

$$\frac{\mathrm{d}i_1}{\mathrm{d}t} = \frac{1}{L_f}(-(R_f + R_d)i_1 + R_d i_2 - u_C + u_0 + u_v) \tag{4.24}$$

on using the relation $u_f = u_v$. The state equation for this plant is the same as (4.20) with suffix v and the matrices almost the same as (4.19):

$$A_v = \begin{pmatrix} -\frac{R_f+R_d}{L_f} & -\frac{1}{L_f} \\ \frac{1}{C_f} & 0 \end{pmatrix}, \quad B_v = (B_{1v}, B_{2v}), \quad B_{1v} = \begin{pmatrix} \frac{R_d}{L_f} & 0 \\ 0 & 0 \end{pmatrix}, \tag{4.25}$$

$$B_{2v} = \begin{pmatrix} \frac{1}{L_f} \\ 0 \end{pmatrix}, \quad C_{1v} = (-R_d, -1), \quad D_{1v} = (R_d, 1), \quad D_{2v} = 0,$$

where A_v is the cofactor $\tilde{a}_{22} = \tilde{a}_{22}$.

4.3 Unification

In this section we shall unify linear systems (4.20) given above. For this we view (4.7) as **impedance operators** $Z = Z(t)$ with current I flowing through it, i.e.

$$Z_L I = -L\frac{\mathrm{d}I}{\mathrm{d}t}, \quad Z_C I = -\int_0^t I\,\mathrm{d}t, \quad Z_R = -RI, \tag{4.26}$$

where L, C, and R indicate the coil, condenser and resistance, respectively.

We consider the cascade connection $\mathcal{Z}$ of two impedances $Z_i = Z_i(t)$, $i = 1, 2$ with the potential u and u_0 and with the current I flowing from u to u_0, thus the voltage difference is $u_0 - u$.

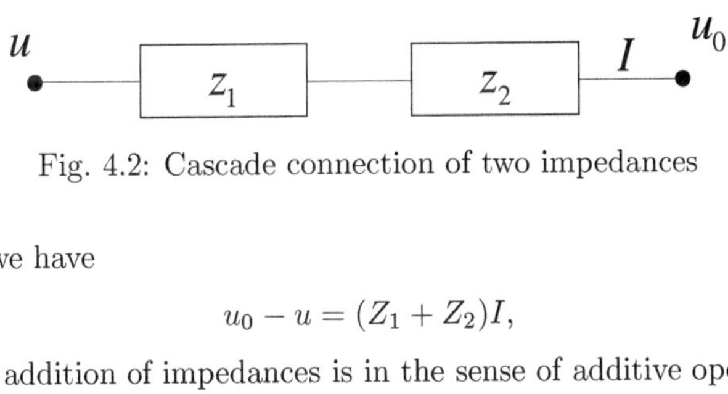

Fig. 4.2: Cascade connection of two impedances

Then we have

$$u_0 - u = (Z_1 + Z_2)I, \tag{4.27}$$

where the addition of impedances is in the sense of additive operators, i.e. $Z_1 I + Z_2 I$.

Suppose Z_2 is a parallel connection of the coil with inductance L and a resistor with resistance r and that the current flowing the coil is i and that Z_1 is a resister with resistance R_1. Then

$$Z_2 I = -L\frac{\mathrm{d}i}{\mathrm{d}t} = -r(I - i), \quad u_0 - u = -R_1 I - r(I - i). \tag{4.28}$$

Substituting the first equality in (4.28) into (4.27), we obtain

$$u_0 - u = -(R_1 + r)I + ri$$

or

$$I = \frac{r}{R_1 + r} i - \frac{1}{R_1 + r}(u_0 - u). \tag{4.29}$$

Substituting this in (4.28), we obtain

$$L\frac{\mathrm{d}i}{\mathrm{d}t} = r\left(\frac{r}{R_1 + r} - 1\right) i - \frac{r}{R_1 + r}(u_0 - u),$$

whence

$$\frac{\mathrm{d}i}{\mathrm{d}t} = -\frac{R_1 r}{(R_1 + r)L} i - \frac{r}{(R_1 + r)L}(u_0 - u). \tag{4.30}$$

More generally, we consider the combination of two such cascade connections at node u_0. Two impedances $Z_i = Z_i(t)$, $i = 3, 4$ with the potential $u = 0$ and u_0 and with the current I_2 flowing from u to u_0. Then we have

$$u_0 = (Z_3 + Z_4)I_2. \tag{4.31}$$

We choose $Z_3 = Z_C = u_C$ (condenser) and Z_4 a resister with resistance R_4. Then (4.31) becomes

$$u_0 = u_C + R_4(I_1 - I_2), \tag{4.32}$$

where I_1 is the current flowing the impedances Z_1, Z_2 in (4.27). Hence substituting (4.32) in (4.30), we conclude

Theorem 4.1. *If two cascade connections of two impedances Z_i, $i = 1, \cdots, 4$ are connected at the node u_0 with voltage difference $u_0 - u$ and i flowing through it, then (4.28) in the form*

$$u_0 - u = -R_1 I - L\frac{\mathrm{d}i}{\mathrm{d}t} \tag{4.33}$$

describes the whole paradigm as either

$$\frac{\mathrm{d}i}{\mathrm{d}t} = -\frac{1}{(R_1 + r)L}(rR_1 i - R_4 I_1) - \frac{R_4}{(R_1 + r)L} I_2 - \frac{1}{(R_1 + r)L}(u_C - u) \tag{4.34}$$

or

$$\frac{\mathrm{d}i}{\mathrm{d}t} = -\frac{R_1 + R_4}{L} I_1 + \frac{R_4}{L} I_2 - \frac{1}{L}(u_C - u), \tag{4.35}$$

where u_C is described in Corollary 4.1.

Corollary 4.1. *The cascade connection (4.30) of two impedances Z_1 (resistance) and Z_2 (the parallel connection of a coil and a resistance) is a special case of (4.34) with $I_1 = I_2$ and (therefore) $u_C = u_0$. (4.12) is a special case with $I_1 = i_1$, $I_2 = i_2$.*

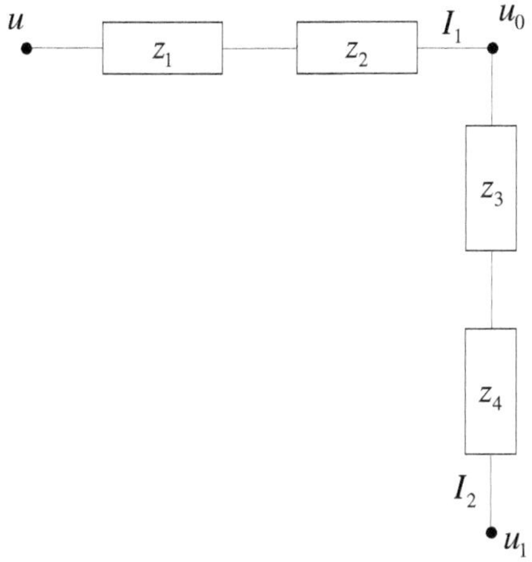

Fig. 4.3: Combination of cascade connections

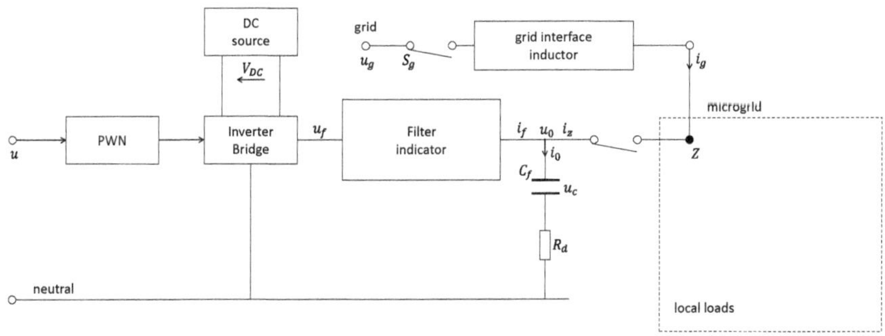

Fig. 4.4: Control plant P in Fig. 4.1 modified

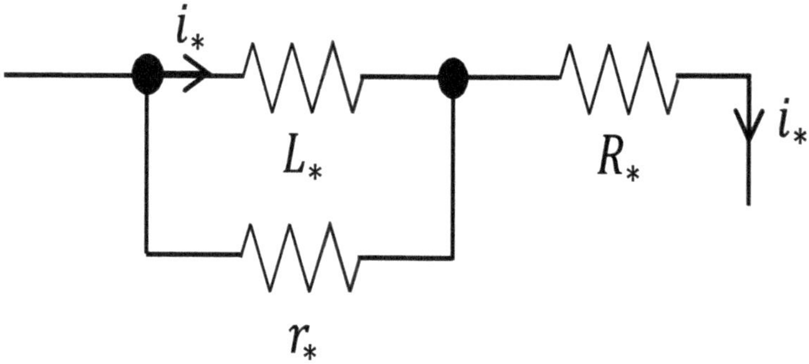

Fig. 4.5: Components of P above

In this figure, $*$ is to be replaced by f if it is a filter, by g if it is a grid and C if it is a conductor. And at the upper right corner is a node, then the current ramifies into i_{*1} and i_{*2}.

Example 4.1. We consider the three cascade connections $\mathcal{Z}_i$, $i = 1, 2, 3$ connected at a node $\mathcal{Z}$ with the flowing-in currents i_f, i_g and flowing-out currents i, i_d (d is for harmonic distortion). And the configurations of each $\mathcal{Z}_i$ are similar. $\mathcal{Z}_1$ indicates the filter inductor with Z_2 a parallel connection of the coil with inductance L_f and a resistor with resistance r_f and that the current flowing the coil is i_1 and that Z_1 is a resister with resistance R_f. Other two are similar as given Table 4.1 below.

In this case, (4.29) reads

$$i_f = \frac{r_f}{R_f + r_f} i_1 - \frac{1}{R_f + r_f}(u_0 - u), \quad i_g = \frac{r_g}{R_g + r_g} i_2 - \frac{1}{R_g + r_g}(u_0 - u_g), \tag{4.36}$$

$$i = \frac{r}{R + r} i_3 + \frac{1}{R + r} u_0.$$

Hence from Theorem 4.1 it follows that

$$\begin{aligned} \frac{\mathrm{d}i_1}{\mathrm{d}t} &= -\frac{R_f r_f}{(R_f + r_f)L_f} i_1 - \frac{r_f}{(R_f + r_f)L_f}(u_0 - u), \\ \frac{\mathrm{d}i_2}{\mathrm{d}t} &= -\frac{R_g r_g}{(R_g + r_g)L_g} i_2 - \frac{r_g}{(R_g + r_g)L_g}(u_0 - u_g), \\ \frac{\mathrm{d}i_3}{\mathrm{d}t} &= -\frac{Rr}{(R + r)L} i_3 + \frac{r}{(R + r)L} u_0. \end{aligned} \tag{4.37}$$

We want to add one more state variable u_0 which is the electro-motive force u_c generated by the condenser C. Since the current flowing the condenser is $i_f - i_0$, we have

$$\frac{\mathrm{d}}{\mathrm{d}t}u_0 = \frac{1}{C}(i_f - i_0). \tag{4.38}$$

At the node $\mathcal{Z}$, we have by the Kirchhoff law,

$$i_0 + i_g = i + i_d,$$

whence

$$i_0 = i + i_d - i_g. \tag{4.39}$$

Hence (4.38) amounts to

$$\frac{\mathrm{d}}{\mathrm{d}t}u_0 = \frac{1}{C}(i_f + i_g - i - i_d). \tag{4.40}$$

Substituting (4.36) in this, we deduce that

$$\begin{aligned}\frac{\mathrm{d}}{\mathrm{d}t}u_0 = \frac{1}{C}\Bigg(&\frac{r_f}{R_f + r_f}i_1 + \frac{r_g}{R_g + r_g}i_2 - \frac{r}{R+r}i_3 \\ &- \left(\frac{r_f}{R_f + r_f} + \frac{r_g}{R_g + r_g} + \frac{r}{R+r}\right)u_0 \\ &- i_d + \frac{1}{R_g + r_g}u_g + \frac{1}{R_f + r_f}u\Bigg).\end{aligned} \tag{4.41}$$

We put

$$\boldsymbol{x} = \begin{pmatrix} i_1 \\ i_2 \\ i_3 \\ u_0 \end{pmatrix}, \quad \boldsymbol{u} = \begin{pmatrix} \boldsymbol{w} \\ u \end{pmatrix}, \quad \boldsymbol{w} = \begin{pmatrix} i_d \\ u_g \\ i_{\mathrm{ref}} \\ u \end{pmatrix}, \quad \boldsymbol{y} = \begin{pmatrix} e \\ i_0 \end{pmatrix}, \tag{4.42}$$

where $e = u_{\mathrm{ref}} - u_0$. Then, putting

$$A = \begin{pmatrix} -\frac{R_f r_f}{(R_f+r_f)L_f} & 0 & 0 & -\frac{r_f}{(R_f+r_f)L_f} \\ 0 & -\frac{R_g r_g}{(R_g+r_g)L_g} & 0 & -\frac{r_g}{(R_g+r_g)L_g} \\ 0 & 0 & -\frac{Rr}{(R+r)L} & -\frac{r}{(R+r)L} \\ \frac{r_f}{R_f+r_f} & \frac{r_g}{R_g+r_g} & -\frac{r}{R+r} & -\frac{1}{C}\left(\frac{r_f}{R_f+r_f}+\frac{r_g}{R_g+r_g}+\frac{r}{R+r}\right) \end{pmatrix}, \tag{4.43}$$

$$B = (B_1, B_2), \quad B_1 = \begin{pmatrix} 0 & 0 & 0 \\ 0 & \frac{r_g}{(R_g+r_g)L_g} & 0 \\ 0 & 0 & 0 \\ \frac{1}{C} & \frac{1}{(R_g+r_g)C} & 0 \end{pmatrix}, \quad B_2 = \begin{pmatrix} \frac{r_f}{(R_f+r_f)L_f} \\ 0 \\ 0 \\ \frac{1}{(R_f+r_f)C} \end{pmatrix},$$

$$C = \begin{pmatrix} C_1 \\ C_2 \end{pmatrix}, C_1 = (0,0,0,-1), C_2 = \left(0, -\frac{r_g}{R_g+r_g}, \frac{r}{R+r}, \frac{1}{R+r} + \frac{1}{R_g+r_g}\right),$$

$$D = \begin{pmatrix} D_{11} & D_{12} \\ D_{21} & D_{22} \end{pmatrix} = \begin{pmatrix} 0 & 0 & 1 & 0 \\ 1 & -\frac{1}{R_g+r_g} & 0 & 0 \end{pmatrix},$$

(4.37) and (4.41) lead to

$$\frac{\mathrm{d}}{\mathrm{d}t}\boldsymbol{x} = A\boldsymbol{x} + B\boldsymbol{u} = A\boldsymbol{x} + (B_1, B_2)\begin{pmatrix} \boldsymbol{w} \\ u \end{pmatrix}, \tag{4.44}$$

$$\boldsymbol{y} = C\boldsymbol{x} + D\boldsymbol{u} = \begin{pmatrix} C_1 \\ C_2 \end{pmatrix}\boldsymbol{x} + \begin{pmatrix} D_{11} & D_{12} \\ D_{21} & D_{22} \end{pmatrix}\begin{pmatrix} \boldsymbol{w} \\ u \end{pmatrix}.$$

connection	coil	c. resist.	c. current	resistance	current
$\mathcal{Z}_1$	L_f	r_f	i_1	R_f	i_f
$\mathcal{Z}_2$	L_g	r_g	i_2	R_g	i_g
$\mathcal{Z}_3$	L	r	i_3	R	i_0

Table 4.1: Components of $\mathcal{Z}_i$'s

The state $\boldsymbol{x}$ is not visible while the input and output are so, and the state may be thought of as an interface between the past and the present information since it contains all the information contained in the system from the past. The $\boldsymbol{x}$ being invisible, (4.1) would read

$$\boldsymbol{y} = D\boldsymbol{u}, \tag{4.45}$$

which appears in many places in literature in disguised form. All the subsequent systems e.g. (4.46) are variations of (4.45). And whenever we would like to obtain the state equation, we are to restore the state $\boldsymbol{x}$ to make a recourse to (4.1), which we would call the **visualization principle**. In the case of feedback system, it is often the case that (4.45) is given in the form of (4.49). It is quite remarkable that this controller S works for the matrix variable in the symplectic geometry (cf. §4.5).

The visualization principle implies the **embedding principle**: Given a state space representation of a transfer function $G(s)$, it is to be embedded in the **state equation** (4.1).

4.4 H^∞-controllers

[Li *et al.* (2012), Chapter 6] introduces the Hardy space H^p which consists of functions $f(s)$ analytic in $\mathcal{RHP}$—right half-plane $\sigma > 0$ and such that $f(j\omega) \in L^p$, where L^p is the Banach space of pth power integrable functions.

Following [Kimura (1997), p. 7, p. 67], we first give the definition of a chain scattering representation of a system.

Suppose $\boldsymbol{z} \in \mathbb{R}^m$, $\boldsymbol{y} \in \mathbb{R}^q$, $\boldsymbol{w} \in \mathbb{R}^r$ and $\boldsymbol{u} \in \mathbb{R}^p$ are related by

$$\begin{pmatrix} \boldsymbol{z} \\ \boldsymbol{y} \end{pmatrix} = P \begin{pmatrix} \boldsymbol{w} \\ \boldsymbol{u} \end{pmatrix}, \tag{4.46}$$

where

$$P = \begin{pmatrix} P_{11} & P_{12} \\ P_{21} & P_{22} \end{pmatrix}. \tag{4.47}$$

According to the embedding principle, this is to be thought of as (4.45) corresponding to the second equality in (4.1).

(4.46) means that

$$\boldsymbol{z} = P_{11}\boldsymbol{w} + P_{12}\boldsymbol{u}, \quad \boldsymbol{y} = P_{21}\boldsymbol{w} + P_{22}\boldsymbol{u}. \tag{4.48}$$

Suppose that $\boldsymbol{y}$ is fed back to $\boldsymbol{u}$ by

$$\boldsymbol{u} = C\boldsymbol{y}, \tag{4.49}$$

where C is a **controller**. Multiplying the second equality in (4.48) by S and incorporating (4.49), we find that

$$\boldsymbol{u} = C\boldsymbol{y} = CP_{21}\boldsymbol{w} + CP_{22}\boldsymbol{u},$$

whence $\boldsymbol{u} = (I - P_{22}K)^{-1}CP_{21}\boldsymbol{w}$.

Let the **closed-loop transfer function** Φ be defined by

$$\boldsymbol{z} = \Phi\boldsymbol{w}. \tag{4.50}$$

Φ is given by

$$\Phi = P_{11} + P_{12}(E - P_{22}C)^{-1}CP_{21}. \tag{4.51}$$

(4.51) is sometimes referred to as a **linear fractional transformation** and denoted by

$$LF(P; K).$$

H^∞-control problem

Find a controller K such that the closed-loop system is internally stable and the transfer function Φ satisfies

$$\|\Phi\|_\infty < \gamma \tag{4.52}$$

for a positive constant γ, where the norm in (4.52) is the H^∞ norm (which may be thought of as the limit $p \to \infty$, [Li *et al.* (2012), Theorem 6.4]).

4.4.1 *Chain scattering representation*

Assume that P_{21} is a *(square) regular* matrix (whence $q = r$). Then from the second equality of (4.48), we obtain

$$\boldsymbol{b}_1 = P_{21}^{-1}(\boldsymbol{a}_2 - P_{22}\boldsymbol{b}_2) = -P_{21}^{-1}P_{22}\boldsymbol{b}_2 + P_{21}^{-1}\boldsymbol{a}_2. \tag{4.53}$$

Substituting (4.53) in the first equality of (4.48), we deduce that

$$\boldsymbol{a}_1 = (P_{12} - P_{11}P_{21}^{-1}P_{22})\boldsymbol{b}_2 + P_{11}P_{21}^{-1}\boldsymbol{a}_2. \tag{4.54}$$

Hence putting

$$\Theta = CHAIN(P) = \begin{pmatrix} P_{12} - P_{11}P_{21}^{-1}P_{22} & P_{11}P_{21}^{-1} \\ -P_{21}^{-1}P_{22} & P_{21}^{-1} \end{pmatrix} \tag{4.55}$$

$$= \begin{pmatrix} \Theta_{11} & \Theta_{12} \\ \Theta_{21} & \Theta_{22} \end{pmatrix},$$

which is usually referred to as a **chain scattering representation** of P, we obtain an equivalent form of (4.46)

$$\begin{pmatrix} \boldsymbol{a}_1 \\ \boldsymbol{b}_1 \end{pmatrix} = CHAIN(P)\begin{pmatrix} \boldsymbol{b}_2 \\ \boldsymbol{a}_2 \end{pmatrix} = \begin{pmatrix} \Theta_{11} & \Theta_{12} \\ \Theta_{21} & \Theta_{22} \end{pmatrix}\begin{pmatrix} \boldsymbol{a}_2 \\ \boldsymbol{b}_2 \end{pmatrix}. \tag{4.56}$$

Substituting (4.49), (4.56) becomes

$$\begin{pmatrix} \boldsymbol{a}_1 \\ \boldsymbol{b}_1 \end{pmatrix} = \begin{pmatrix} \Theta_{11}S + \Theta_{12} \\ \Theta_{21}S + \Theta_{22} \end{pmatrix}\boldsymbol{a}_2,$$

whence we deduce that

$$\Phi = (\Theta_{11}S + \Theta_{12})(\Theta_{21}S + \Theta_{22})^{-1} = \Theta S, \tag{4.57}$$

the **linear fractional transformation** (which is referred to as a **homographic transformation** and denoted by $HM(\Phi; S)$), where in the last equality we mean the action of Θ on the variable S. We must impose the non-constant condition $|\Theta| \neq 0$. Then $\Theta \in GL_{m+r}(\mathbb{R})$.

If S is obtained from S' under the action of Θ', $S = \Theta' S'$, then its composition J with (4.57) yields $JS' = \Phi\Phi' = \Theta\Theta' S'$, i.e.

$$J = \Theta\Theta', \quad HM(\Theta; HM(\Theta'; S)) = HM(\Theta\Theta'; S), \tag{4.58}$$

which is referred to as the **cascade connection** or the cascade structure of Θ and Θ'.

Thus *the chain-scattering representation of a system allows us to treat the feedback connection as a cascade connection.*

The restriction that P_{21} is to be a square matrix has been removed in [Xin and Kimura (1994a)] and [Xin and Kimura (1994b)].

system	functions	action	region of convergence	critical line
S	rational	symplectic	$\sigma > 0$	$\sigma = 1$
ζ	meromorphic	modular	$\sigma > 1$	$\sigma = \frac{1}{2}$

Table 4.2: Correspondence between control systems and the Riemann zeta-function $\zeta(s)$ in (3.129)

4.5 Siegel upper space

Let $*$ denote the conjugate transpose of a square matrix: $S^* = {}^t\bar{S}$ and let the imaginary part of S defined by $\operatorname{Im} S = \frac{1}{2j}(S - S^*)$. Let $\mathcal{H}_n$ be the **Siegel upper half-space** consisting of all the matrices S (recall Eq. (4.49)) whose imaginary parts are positive definite ($\operatorname{Im} S > 0$—imaginary parts of all eigen-values are positive) and satisfies $S = {}^tS$:

$$\mathcal{H}_n = \{S \in \mathrm{M_n}(\mathbb{C}) \,|\, \operatorname{Im} S > 0, \ S = {}^tS\} \tag{4.59}$$

and let $\mathrm{Sp}(n, \mathbb{R})$ denote the **symplectic group** of order n:

$$\mathrm{Sp}(n, \mathbb{R}) \tag{4.60}$$
$$= \left\{ \Theta = \begin{pmatrix} \Theta_{11} & \Theta_{12} \\ \Theta_{21} & \Theta_{22} \end{pmatrix} \,\middle|\, \begin{pmatrix} \Theta_{11} & \Theta_{12} \\ \Theta_{21} & \Theta_{22} \end{pmatrix}^{-1} = \begin{pmatrix} \Theta_{22} & -{}^t\Theta_{12} \\ -{}^t\Theta_{21} & \Theta_{11} \end{pmatrix} \right\}.$$

The **action** of $\mathrm{Sp}(n, \mathbb{R})$ on $\mathcal{H}_n$ is defined by (4.57) which we restate as

$$\Theta S = (\Theta_{11} S + \Theta_{12})(\Theta_{21} S + \Theta_{22})^{-1} (= \Phi). \tag{4.61}$$

Theorem 4.2. *For a controller S living in the Siegel upper space, its rotation $Z = -jS$ lies in the right half-space $\mathcal{RHS}$, i.e. stable having positive real parts. For the controller Z, the feedback connection*

$$-j\boldsymbol{b}_2 = Z(-j\boldsymbol{a}_2) \tag{4.62}$$

is accommodated in the cascade connection of the chain scattering representation Θ (4.58), which is then viewed as the action (4.58) of $\Theta \in \mathrm{Sp}(n, \mathbb{R})$ on $S \in \mathcal{H}_n$:

$$(\Theta\Theta')S = \Theta(\Theta' S); \quad or \quad HM(\Theta; HM(\Theta'; S)) = HM(\Theta\Theta'; S), \tag{4.63}$$

where Θ is subject to the condition

$${}^t\bar{\Theta}U\Theta = U, \tag{4.64}$$

with $U = \begin{pmatrix} O & I_n \\ -I_n & O \end{pmatrix}$, An FOPID controller (in §4.7), being a unity feedback connection, is also accommodated in this framework.

Remark 4.1. With action, we may introduce the orbit decomposition of $\mathcal{H}_n$ and whence the fundamental domain. We note that in the special case of $n = 1$, we have $\mathcal{H}_1 = \mathcal{H}$ and $\mathrm{Sp}(1, \mathbb{R}) = \mathrm{SL}_n(\mathbb{R})$ and the theory of modular forms of one variable is well-known. Siegel modular forms are a generalization of the one variable case into several variables. As in the case of the sushmna principle in [Kanemitsu and Tsukada (2007)], there is a need to rotate the upper half-space into the right half-space $\mathcal{RHS}$, which is a counterpart of the right-half-plane $\mathcal{RHP}$. In the case of Siegel modular forms, the matrices are constant ones, while in control theory, they are analytic functions (mostly rational functions analytic in $\mathcal{RHP}$). A general theory would be useful for control theory. There are many research problems lying in this direction.

4.6 Classification of signals

This section is an extract from [Li *et al.* (2017), §6.5] with some modifications and additions. In mathematics and in particular in analytic number theory, studying the mean square or higher power moments in the form of a sum or an integral is quite common. Especially, this idea is applied to finding out the true order of magnitude of the error term on average. Such an average result will give a hint on the order of the error term itself. In control theory this type of averages are used to minimize the errors, e.g. in §4.8.3.

Recall the Banach space L^p of pth power integrable functions $f(t)$ in §4.4. Although this is a great warp from Riemann integration, it has a limitation that $\|f(t)\| \to 0$ as $t \to \infty$ fot the p norm to be finite, excluding signals of infinite duration such as unit step signals or periodic ones. To circumvent the inconvenience, the notion of averaged norm, or the mean $2k$th power value

$$M_{2k}(f) = M_{2k}(f,T) = \frac{1}{T}\int_0^T \|f(t)\|^{2k}\,\mathrm{d}t \tag{4.65}$$

or similar, has been introduced and intensively studied both in analytic number theory and control theory. In the latter the **power norm** is often used with $k = 1$:

$$\mathrm{power}_{2k}(f) = \lim_{T\to\infty} M_{2k}(f,T)^{\frac{1}{2k}} = \lim_{T\to\infty}\left(\frac{1}{T}\int_0^T \|f(t)\|^{2k}\,\mathrm{d}t\right)^{\frac{1}{2k}}. \tag{4.66}$$

Let $\zeta(s)$ denote the Riemann zeta-function defined by (3.129) having a simple pole at $s = 1$. It is essential that it does not vanish on the line $\sigma = 1$ for the prime number theorem (PNT; cf. e.g. [Li *et al.* (2017), p. 80 (3.35)]) to hold. The plausible best bound for the error term for the PNT is equivalent to the celebrated **Riemann hypothesis (RH)** to the effect that the Riemann zeta-function does not vanish on the critical line $\sigma = \frac{1}{2}$. Since the values on the critical line are expected to be small, the averaged norm $M_2(\zeta)$ or $M_4(\zeta)$, i.e. the mean value

$$M_{2k}(\zeta,\sigma) = \frac{1}{T}\int_0^T \left|\zeta\left(\sigma+it\right)\right|^{2k}\,\mathrm{d}t \tag{4.67}$$

for $\sigma \ge \frac{1}{2}$, $k \ge 1$ is of great interest and there have appeared a great deal of research work on the subject. The first result for $M_4\left(\zeta,\frac{1}{2}\right)$ is due to Ingham who used the approximate functional equation for the Riemann zeta-function to obtain (4.70).

Proposition 4.1. *Let $d_k(n)$ denote the Piltz divisor function (k-fold divisor function) generated by*

$$\zeta^k(s) = \sum_{n=1}^{\infty}\frac{d_k(n)}{n^{2s}}, \quad \sigma > 1. \tag{4.68}$$

Then as $T \to \infty$ we have

$$M_{2k}(\zeta,\sigma) = \frac{1}{T}\int_0^T \left|\zeta\left(\sigma+it\right)\right|^{2k}\,\mathrm{d}t = \sum_{n=1}^{\infty}\frac{d_k^2(n)}{n^{2\sigma}}(1+o(1)) \tag{4.69}$$

for $\sigma > 1 - \frac{1}{k}$ and every integer $k > 2$ while

$$M_4\left(\zeta,\frac{1}{2}\right) = \frac{1}{T}\int_0^T \left|\zeta\left(\frac{1}{2}+it\right)\right|^{4}\,\mathrm{d}t = \frac{1}{4\pi^2}\log^4 T(1+o(1)). \tag{4.70}$$

See e.g. [Bellman (1980)]. The main interest in such estimates as (4.70) lies in the fact that estimates for all $k \in \mathbb{N}$

$$M_{2k}\left(\zeta, \frac{1}{2}\right) = O(T^a) \tag{4.71}$$

would imply the weak Lindelöf hypothesis (LH) in the form

$$\zeta\left(\frac{1}{2} + it\right) = O(T^{\frac{a}{2k}+\varepsilon}) \tag{4.72}$$

for every $\varepsilon > 0$. It is apparent that the RH implies the LH.

Cf. [Ivić (1985)] and [Titchmarsh (1986)] for more details on power moments.

The class of functions f with power norm $\text{power}_2(f) = 0$, called the class of **finite energy signals**, has been extensively studied and which clearly contains the subclass L^2, which we may call **finite energy functions**. As is mentioned in Papoulis [Papoulis (1962)], it would be worth considering the class of functions with non-zero average power, i.e. $0 < \text{power}_{2k}(f) < \infty$, the **finite power signals** class. Such functions do not have Fourier transforms in general. An important feature is that this class contains all periodic functions. Indeed, let $f(t)$ be a periodic function given by the **Fourier series**

$$f(t) = \sum_{n=-\infty}^{\infty} a_n e^{in\omega_0 t} \tag{4.73}$$

with $\sum_{n=-\infty}^{\infty} |a_n|^2 < \infty$, so that $f \in \ell^2$. Cf. §8.2.1 for basics of Fourier analyis. Then

$$\text{power}_{2k}(f) = \frac{2\pi}{\omega_0} \int_{-\frac{\pi}{\omega_0}}^{\frac{\pi}{\omega_0}} |f(t)|^2 \, \mathrm{d}t = \sum_{n=-\infty}^{\infty} |a_n|^2. \tag{4.74}$$

The class of finite power signals also contains non-periodic function which are usually specified by some averages. The most common one is the **autocorrelation** defined by

$$R_f(t) = \lim_{T\to\infty} \frac{1}{2T} \int_{-T}^{T} \bar{f}(\tau) f(t+\tau) \, \mathrm{d}\tau. \tag{4.75}$$

Proposition 4.1 implies that the Riemann zeta-function has positive power norm value for $\sigma > \frac{1}{2}$ so that it is a finite power signal for $\sigma > \frac{1}{2}$ and that $\zeta\left(\frac{1}{2} + it\right)$ is not a finite but an **infinite power signal**. This suggests the formidable difficulty of the LH, and *a fortiori* the RH.

signal	finite-energy functions	finite power signal	infinite power signal
power	$\mathrm{power}_2(f) = 0$	$0 < \mathrm{power}_{2k}(f) < \infty$	$\mathrm{power}_{2k}(f) = \infty$
example	periodic functions	$\zeta(1+it)$	$\zeta\left(\frac{1}{2}+it\right)$

Table 4.3: Classification of signals

We recall that the transfer function defined by (4.5) is usually a simple rational function (proper fraction) in s, so that it lies in the space L^p and is far simpler than number-theoretic zeta-functions, which are finite or even infinite power signals. It seems appropriate to study control theory as that of finite entity such as a fraction or a polynomial. Therefore it would be promising to work with Dirichlet polynomials as in [Huxley (1972), p. 84] or from the point of view of algebraic geometry as in [Falb (1990)] and [Falb (2000)], thus a new approach "finite approaches to control theory" is in order.

4.7 FOPID

"FO" means "Fractional order" and "PID" refers to "Proportional, Integral, Differential", where "Proportional" means just constant times the input function $e(t)$, "Integral" means the **fractional order integration** $I_t^\lambda D_t^{-\lambda}$ of $e(t)$ $(\lambda > 0)$, and "Differential" the **fractional order differentiation** D_t^δ of $e(t)$ $(\delta > 0)$.

The FO $PI^\lambda D^\delta$ controller (control signal in the time domain) is one of the most refined feed-forward compensator defined by

$$u(t) = \left(K_p + K_i D_t^{-\lambda} + K_d D_t^\delta\right) e(t), \tag{4.76}$$

where u is the input function, e is the deviation and K_p, K_i, K_d are constant parameters which are to be specified (K_p- the position feedback gain, K_d- the velocity feedback gain). DE (4.76) translates into the **state equation**

$$Y(s) = C(s)E(s), \tag{4.77}$$

where U, Y indicate the Laplace transforms of u, y, respectively and G is the compensator **continuous transfer function**

$$C(s) = K_p + K_i s^{-\lambda} + K_d s^\delta. \tag{4.78}$$

The derivation of (4.78) from (4.76) depends on the following. The general fractional calculus operator ${}_aD_t^\alpha$ is symbolically stated as

$${}_aD_t^\alpha = \begin{cases} \frac{\mathrm{d}^\alpha}{\mathrm{d}t^\alpha}, & \mathrm{Re}\,\alpha > 0 \\ 1, & \mathrm{Re}\,\alpha = 0 \\ \int_a^t \frac{1}{\mathrm{d}t^\alpha}, & \mathrm{Re}\,\alpha < 0, \end{cases} \tag{4.79}$$

where a and t are the lower and upper limits of integration and α is the order of calculus.

More precisely, the definition of the fractional differointegral is given by the Riemann-Liouville expression

$$ {}_aD_t^\alpha f(t) = \frac{1}{\Gamma(1-\{\alpha\})}\left(\frac{\mathrm{d}}{\mathrm{d}t}\right)^{\alpha-\{\alpha\}+1}\int_a^t (t-\tau)^{-\{\alpha\}} f(\tau)\,\mathrm{d}\tau \tag{4.80} $$

where $\{\alpha\} = \alpha - [\alpha]$ indicates the fractional part of α, with $[\alpha]$ the integral part of α. Thus we are also led to the Riemann-Liouville fractional integral transform:

$$ \mathcal{RL}[f] = \frac{1}{\Gamma(\mu)}\int_0^y (y-x)^{\mu-1} f(x)\,\mathrm{d}x. \tag{4.81} $$

When $\alpha \in \mathbb{N}$, (4.80) reads

$$ {}_aD_t^\alpha f(t) = \left(\frac{\mathrm{d}}{\mathrm{d}t}\right)^{\alpha+1}\int_a^t f(\tau)\,\mathrm{d}\tau = f^{(\alpha)}(t), \tag{4.82} $$

the αth derivative of f.

We shall see that the definition (4.80) is a natural outcome of the general formula for the difference operator of order $\alpha \in \mathbb{N}$ with difference $y \geq 0$:

$$ \Delta_y^\alpha f(x) = \sum_{\nu=0}^{\alpha} (-1)^{\alpha-\nu}\binom{\alpha}{\nu} f(x+\nu y). \tag{4.83} $$

If f has the αth derivative $f^{(\alpha)}$, then

$$ \Delta_y^\alpha f(x) = \int_x^{x+y} \mathrm{d}t_1 \int_{t_1}^{t_1+y} \mathrm{d}t_2 \cdots \int_{t_{\alpha-1}}^{t_{\alpha-1}+y} f^{(\alpha)}(t_\alpha)\,\mathrm{d}t_\alpha. \tag{4.84} $$

The special case of (4.84) with $t_\nu = a, a+y \to x$ ($\varphi(t) = f^{(\alpha)}(t)$) reads

$$ \Delta_{x-a}^\alpha \varphi(x) = \int_a^x \mathrm{d}t \int_a^x \mathrm{d}t \cdots \int_a^x \varphi(t)\,\mathrm{d}t = \frac{1}{\Gamma(\alpha)}\int_a^x (x-t)^{\alpha-1}\varphi(t)\,\mathrm{d}t, \tag{4.85} $$

whose far-right hand side is $\mathcal{RL}[\varphi]$.

Let $F(s)$ be the Laplace transform of the input function $f(t)$. Then

$$ L[{}_0D_t^\alpha f](t) = s^\alpha F(s) - {}_0D_t^{\alpha-1} f(t)\,|_{t=0} \tag{4.86} $$

and

$$ L[{}_0D_t^{-\alpha} f](t) = s^{-\alpha} F(s). \tag{4.87} $$

4.8 PSO optimization for balanced charge of Li-ion batteries

4.8.1 *Introduction*

In recent times, more and more need arose to effectively charge and discharge Lithium-ion batteries, Li-ion batteries for short.

In electric motor drives for automobile applications (esp. EV's), there is a need of DC supply current in the range of 450 V. This is achieved by a *DC–DC* converter which links the battery stack and motor drive. For its optimal operation, the stack voltage of 150 V–300 V is required.

Although recent Li-ion batteries or super-capacitors have amazing storage capacities, a cell alone will never suffice since the voltage of a common cell of a Li-ion battery is typically 3.3∼3.6. Hence they must be combined to form a block. An example is a series (cascade) connection of 12 cells combined in a block as a 3 parallel array: $3P12S$. This leads to the block voltage of 30∼45 in ideal cases. To meet the above-mentioned required stack voltage, 5 to 8 blocks are combined in a series to form a stack.

Here the principle applies to the effect that a chain is as weak as its weakest ring. I.e. the block voltage is as strong as its lowest voltage of the weakest cell.

In [Bonfiglio and Roessler (2009)], a new method is introduced of using a transformer in order to balanced charging of battery cells. Each nano phosphate type battery cell has the lower and upper limit of 2 V to 3.6 V and beyond this limit, the cell may be damaged.

The main issue here is to attain the maximum value of the stack voltage by optimizing each cell, which can be naturally seen as a problem in the PSO algorithm. I.e. each cell is a particle and its velocity is the maximum value 3.6 V. Or it may be that there are two series connections, series of blocks as well as series of cells, we may need to think out a nested PSO.

Also another issue would be the control of heat while charging.

4.8.2 *PSO*

PSO, **Particle Swarm Optimization**, is a biologically-inspired stochastic optimization technique arising from the family of evolutionary computation and gives better solutions than GA (**Genetic Algorithm**). In the original version of PSO due to Kennedy and Eberhart [Kennedy and Eberhard (1995)], a swarm consists of N particles moving around a prescribed D-dimensional search space $\mathbb{R}^D$. The ith particle is denoted by

$\boldsymbol{x}_i(t) = (x_{i1}, \cdots, x_{iD}), 1 \leq k \leq N$ whose best previous solution (*pbest*) is denoted by $P_i = (p_{i1}, \cdots, p_{iD})$, while the best solution (*gbest*) achieved by the whole swarm is denoted by $P_g = (p_{g1}, \cdots, p_{gD})$. The current velocity (rate of change of its position) of the ith particle is denoted by $\boldsymbol{v}_i(t) = (v_{i1}, \cdots, v_{iD})$.

At each step, each particle moves toward the (particle best) *pbest* and (global best) *gbest* locations. In the improved PSO, particles are manipulated by the following equations

$$\begin{aligned} v_{id}(k+1) &= wv_{id}(k) + c_1\mathfrak{R}_1 (p_{id} - x_{id}(k)) + c_2\mathfrak{R}_2 (p_{gd} - x_{id}), \\ x_{id}(k+1) &= (1 - mc) \cdot x_{id}(k) + mc \cdot v_{id}(k+1), \\ &\quad 1 \leq i \leq N, 1 \leq d \leq D, \end{aligned} \tag{4.88}$$

where c_1 resp. c_2 are positive constants called **cognitive learning rate** resp. **social learning rate**, w is a **time decreasing inertia factor (weight)**, $\mathfrak{R}_j$ is a **random function** with values in $[0, 1]$, and mc is a **momentum factor**. The velocity of the particles is limited to $[V_{\min}, V_{\max}]$ and $V_{\min} = X_{\min}, V_{\max} = X_{\max}$.

A swarm is the battery stack consisting of $N = 180{\sim}216$ particles and a particle searches in the $\mathbb{R}^D$ with $D = 36$. The ith particle is denoted by $\boldsymbol{x}_k(t) = (x_{k1}(t), \cdots, x_{kD}(t)), 1 \leq k \leq N$, and $x_{kd}(t)$ indicates the amount of charge at time t. The velocity of each particle is the difference between the charged amount and the maximum. The voltage limits of each cell is 2 V to 3.6 V. Hence $[V_{\min} = 2, V_{\max} = 3.6]$.

We consider the stochastic process

$$\boldsymbol{x}_i(k+1) = (1 - mc)\boldsymbol{x}_i(k) + mc\boldsymbol{v}_i(k+1), \tag{4.89}$$

or

$$\boldsymbol{x}_i(k+1) = (1 - mc - \Lambda_1\mathfrak{R}_1 - \Lambda_2\mathfrak{R}_2\boldsymbol{x}_i(k) + (mc + w)\boldsymbol{v}_i(k+1) + C_i, \tag{4.90}$$

where

$$C_i = \Lambda_1\mathfrak{R}_1 P_{best,i} - \Lambda_2\mathfrak{R}_2 G_{best}.$$

4.8.3 *Optimization criteria and a fitness function*

Two optimization criteria are often used ITAE and ISE, which are shorthands for *integral of time-weighted absolute error* and *integral of squared error* as opposed to the robust control (4.52):

$$J_1 = \int_0^\infty |e(t)| \, \mathrm{d}t, \quad J_2 = \int_0^\infty e(t)^2 \, \mathrm{d}t. \tag{4.91}$$

J_1 is feasible to good response but its selection performance is not good, while J_1 can track errors quickly but easily lead to oscillation. In the paper, the weighted combination of ITAE and the (square of) control input:

$$J = \int_0^{\infty} \left(w_1|e(t)| + w_2 u^2(t)\right) dt, \tag{4.92}$$

where $0 < w_i < 1$ ($w_1 = 0.99$, $w_2 = 0.001$ is used).

It seems that the fact that PSO is a stochastic process is not seriously considered. We need to explore this side.

4.8.4 *Nested PSO*

In the problem considered by us, we have a stack of 5–8 blocks connected in series. So now these blocks will be the particles in our swarm. The problem here is that the position and the velocity of each block is not known. The block consists of 3–4 parallel arrays of a connection of 12 cells connected in series. Now the position values of these cells inside a block define the position vector of the block and their individual position values are itself governed by the PSO algorithm in the swarm of cells. So basically the PSO algorithm has been nested here. First we need to apply the algorithm for the cells connected in a series inside a block so as to be able to define the position vector for the block. When we have the position vector of the components of the block, then we apply the algorithm on the block with these values of the position of its constituents.

A swarm consists of N particles moving around a prescribed D-dimensional search space $\mathbb{R}^D$. Also there are in all r levels of inter-dependence i.e. for r different applications of PSO algorithm, the input fed is an outcome of another PSO algorithm. So the process will be to apply the standard PSO on the base where the input parameters are known, then its output is to be fed into the next level which is continued r times. The ith particle at the jth level of dependence is denoted by $\boldsymbol{x}_i^j(t) = \left(x_{i1}^{j-1}(t), \cdots, x_{iD}^{j-1}(t)\right), 1 \le k \le N, 1 \le j \le r$ and $x_i^0 = x_i$ whose best previous solution (*pbest*) is denoted by $P_i^j = \left(p_{i1}^{j-1}, \cdots, p_{iD}^{j-1}\right)$, while the best solution (*gbest*) achieved by the whole swarm is denoted by $P_g^j = \left(p_{g1}^{j-1}, \cdots, p_{gD}^{j-1}\right)$. The current velocity (rate of change of its position) of the ith particle is denoted by $\boldsymbol{v}_i^j(t) = \left(v_{i1}^{j-1}, \cdots, v_{iD}^{j-1}\right)$.

So basically this is the whole difference in nested PSO and the normal PSO. The rest of the equations to be applied will be the same with the only difference that at each instant of time, normal PSO will have to be applied r times.

4.8.5 *Implementation*

PSO algorithm needs to predefine some numerical coefficients like the inertia factor, the maximum velocity, the range of positions, swarm size and topology. The ability of global optimization relies on the values of these sets of parameters. In our problem setting, the swarm has 50 birds/blocks each minimizing the value of fitness independently. Each block contains 36 batteries which can be called as the dimension of the problem. The parameters c_1 and c_2 were taken to be 0.5 each. The inertia factor (w) was taken to be 0.6. The $\mathfrak{R}_j$ were the random function, so taken through the built-in rand function of Matlab. With these values, the velocity and position were iterated using equation (4.88) through which we were able to reach the optimum value of position in about 30 iterations for all the elements in the particle.

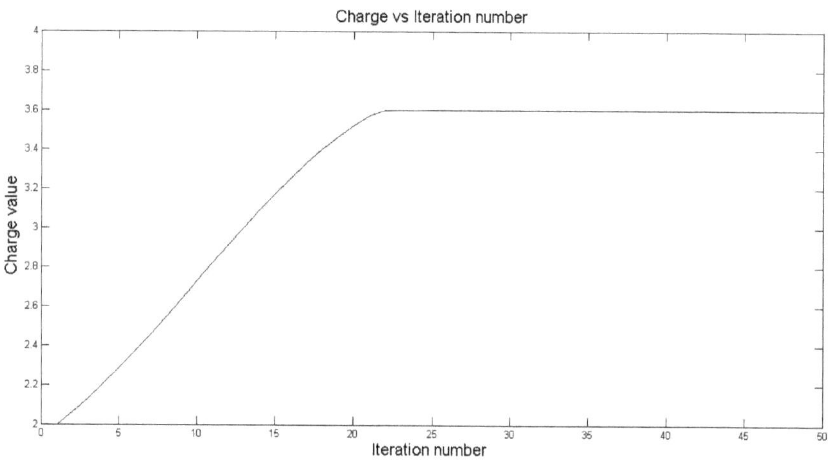

Fig. 4.6: Change of charge of battery vs iteration number

4.8.6 *Fitness value*

The **fitness function** is the function which helps to give an idea to the particle of the direction to which it should move. In our case, the error in position was 3.6 V minus the current value of position for each dimension. We can add the errors of each of the dimension to get the total error of a particle. We simulated with this error function but found that the amount of progress in each iteration is not as good as when we take the maximum of these individual errors. The main reason for maximum being a better judge is that it tells the amount of difference from 3.6 (maximum possible value) on the other hand the sum would vary between 0 and $129.6 = 3.6 \cdot 36$. Now if we get the sum to be some value, we do not know about the number of batteries over which this sum is divided upon. On the other hand if we keep maximum value as the output, it is bound to reduce as we reach the final solution. The figure above shows the variation of position of a dimension (battery) from initially 2 V to the maximum value i.e. 3.6 V. The change of fitness value is shown in the figure below.

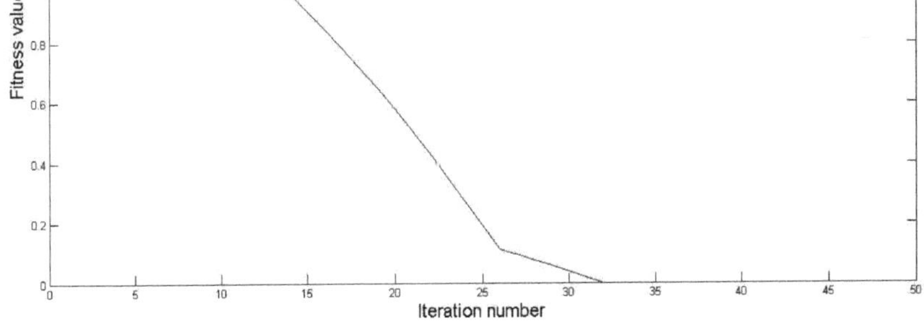

Fig. 4.7: Change of fitness value vs iteration number

Fig. 4.8: Claude Elwood Shannon 1916-2001

Chapter 5

Aspects of chemistry: Adaptation principle

Abstract

In this chapter we treat various topics from chemistry from the point of view of the adaptation principle—things of similar form have similar properties. We shall state the menace of exposure to toxins and show ways to avoid them as much as possible. Toxins include radiation, unhealthy lipids, active oxygen, fertilizers, etc. We also mention the transmigration of some fundamental elements including carbons and nitrogens, the former being responsible for global warming and is discussed in Chapter 6.

5.1 First-order reaction and radioactive decay

Definition 5.1. In an irreversible **first-order reaction** $A \to B$, where the reacting substance A changes into the resulting substance B, we denote by $[A] = [A](t)$, $[B] = [B](t)$ the molarity of A, B, respectively. Then the **rate of increase** $-\frac{d[A]}{dt}$ of B obeys the (first-order) **reaction equation**

$$-\frac{\mathrm{d}[A]}{\mathrm{d}t} = k[A], \tag{5.1}$$

where k is a positive constant called the **reacting constant**, which is determined by experiments. By Proposition 1.1, the solution is given by

$$[A] = [A](t) = e^{-kt}[A](0). \tag{5.2}$$

Example 5.1. In the case of a **radioactive decay**, we write $N = N(t)$ for the number of atoms in the radioactive substance in place of molarity in Definition 5.1 to rewrite (5.1) as

$$\frac{\mathrm{d}N}{\mathrm{d}t} = -kN, \tag{5.3}$$

where k is a positive constant called the **decay constant**. The solution is

$$N = N(t) = e^{-kt}N_0 = e^{-kt}N(0). \tag{5.4}$$

The period of time T needed for the number of atoms to become half of the initial number $N(0) = N_0$ is called a **half-life** (half-value period). From $\frac{1}{2}N_0 = N_0 e^{-kT}$ we deduce the important relation between the decay constant and the half-life:

$$e^{-kT} = \frac{1}{2} \tag{5.5}$$

or

$$T = \frac{\log 2}{k} = \frac{0.693147\cdots}{k}. \tag{5.6}$$

Remark 5.1.

(i) Half-life means that the amount of the substance becomes half and there still remains half. In general, if we denote the period of time needed for the substance decreases to $1/a$ of the initial amount, then (5.6) reads

$$T_{1/a} = \frac{\log a}{k}. \tag{5.7}$$

Hence

$$T_{1/2a} = \frac{\log 2 + \log a}{k} = \frac{0.693147 + \log a}{k}. \tag{5.8}$$

In particular, $T_{1/4} = 2\frac{\log 2}{k} = 2T_{1/2}$, $T_{1/8} = 3\frac{\log 2}{k} = 3T_{1/2}$, meaning that the $1/4$th period is twice as much as the half-life and the $1/8$th period is three times as much. E.g. in the case of cesium in Example 5.2, (ii), it will take 90 years for it to decrease to $1/8$th amount of the initial amount.

(ii) In literature one sometimes finds another formula called the rate of reaction formula, which follows by eliminating the reaction constant from (5.4) by the half-life. Expressing the reaction constant from (5.6),

$$k = \frac{\log 2}{T}. \tag{5.9}$$

Substituting this in (5.4), one obtains

$$N(t) = N_0 e^{-\frac{t \log 2}{T}} = N(0)\left(\frac{1}{2}\right)^{t/T}. \tag{5.10}$$

Example 5.2.

(i) Iodine-131 decays into stable Xe-131: ${}^{131}\mathrm{I} \to {}^{131}\mathrm{Xe}$. The decay constant of ${}^{131}\mathrm{I}$ is 0.00866 (day^{-1}). Hence

$$T = T_{\mathrm{I}} = \frac{0.693147\cdots}{0.00866} = 80.04\cdots(\text{days}).$$

(ii) Cesium 137 (Cs-137) decays into semi-stable Ba − 137m: $^{137}\text{Cs} \to {}^{137m}\text{Ba}$. The decay constant of Cs is $k = 2.31 \times 10^{-2}$ (sec^{-1}). We find half-life of Cs. (5.6) amounts to

$$T = T_{Cs} = \frac{0.693147\cdots}{2.31} \times 10^2 = 30.00\cdots \text{(years)}.$$

(iii) Strontium 90 (^{90}Sr) decays into yttrium ^{90}Y: $^{90}\text{Sr} \to {}^{90}\text{Y}$.
The decay constant of Sr-90 is $k = 7.85 \times 10^{-10}$ (sec^{-1}). We find half-life of Sr-90. (5.6) amounts to

$$T = T_{Sr} = \frac{0.693147\cdots}{7.85} \times 10^{10} = 0.0882862\cdots \times 10^{10} \text{ (sec)}.$$

In years, this is $27.9740\cdots$, after dividing by 3153600.

(iv) In the case of plutonium 239 (^{239}Pu). The decay constant of Pu-239 is $k = 2.88 \times 10^{-5}$ (year^{-1}). We find half-life of Pu. (5.6) amounts to

$$T = T_{Pu} = \frac{0.693147\cdots}{2.88} \times 10^5 = 24,100 \text{ (years)}.$$

(v) Radon Rn = Rn-222 decays into Polonium Po = Po-218: $^{222}\text{Rn} \to {}^{218}\text{Po}$. The reaction continues and there appear four kinds of radioactive substances. The decay constant for Rn is $k = 1.81 \times 10^{-1}$ (day^{-1}). We find half-life of Rn. (5.6) amounts to

$$T = T_{Rn} = \frac{0.693147\cdots}{1.81} \times 10 = 3.825\cdots \text{(days)}.$$

Hence in 4 days, the amount becomes halved.

(vi) (5.9) may be used to compute the decay constant k from half-life. E.g. consider the case of yttrium ^{90}Y as a continuation of (iii) above. Yttrium−90 decays into zirconium−90; $^{90}\text{Y} \to {}^{90}\text{Zr}$. Its half-life is 64 hours. Hence the decay constant is

$$k = k_{\text{Y}} = \frac{0.693147\cdots}{64} = 1.08 \times 10^{-2} \text{ (hour)}^{-1}. \qquad (5.11)$$

Items (iii) and (v) continue on to Example 5.6 below.

It is essential that one is not supposed to remember such a relation but is supposed to be able to derive it from the original reaction equation by integration.

Carbon 14 is a radioactive isotope of the carbon atom which undergoes β decay and changes into Nitrogen 14 by releasing electrons and anti-electronic nutrino with radiating beta rays. Carbon 14 is a threshold of unstable and stable carbon atoms and has been used extensively for age determination—the Carbon 14 method.

$$^{14}\mathrm{C} \to {}^{14}\mathrm{N} + e^- + \bar{\nu}_e. \tag{5.12}$$

Nitrogen 14 is a stable isotope of the Nitrogen atom which occupies 99.63 % of all isotopes.

The half-life of ^{14}C is known to be about 5730 years. Hence as in Example 5.2, (vi), we may compute the decay constant as follows. $k = \frac{0.693147\cdots}{5730} = 1.08 \times 10^{-2}\ (\mathrm{year})^{-1}$.

5.2 Chain reactions

Example 5.3. We use Proposition 1.1 and Theorem 1.10 to solve the system of DEs with initial values $x(0)$, $y(0)$ being given.

$$\begin{cases} -\frac{\mathrm{d}x}{\mathrm{d}t} = k_1 x \\ -\frac{\mathrm{d}y}{\mathrm{d}t} = k_2 y - k_3 x, \end{cases} \tag{5.13}$$

where k_1, k_2, k_3 are constants.

The solution to the first equation is given by (1.246):

$$x = x(t) = e^{-k_1 t} x(0).$$

Hence we may now apply (1.249) with $u = e^{-k_1 t} x(0)$, $a = -k_2$, $b = k_3$ to obtain

$$y = e^{-k_2 t} y(0) + k_3 e^{-k_2 t} x(0) \int_0^t e^{-(k_1 - k_2)\tau}\, \mathrm{d}\tau.$$

Hence if $k_1 \neq k_2$, then

$$y = y(t) = e^{-k_2 t} y(0) + \frac{k_3 x(0)}{k_2 - k_1} \left(e^{-k_1 t} - e^{-k_2 t}\right). \tag{5.14}$$

If $k_1 = k_2$, we have

$$y = (k_3 x(0) t + y(0)) e^{-k_2 t}. \tag{5.15}$$

Exercise 19. Solve (5.13) following the proof of Theorem 1.10.

Solution. The second becomes

$$\frac{\mathrm{d}y}{\mathrm{d}t} + k_2 y = k_3 x,$$

whence noting that

$$\frac{\mathrm{d}}{\mathrm{d}t}(ye^{k_2 t}) = e^{k_2 t}\left(\frac{\mathrm{d}y}{\mathrm{d}t} + k_2 y\right) = e^{k_2 t} k_3 x = k_3 x(0) e^{(k_2-k_1)t},$$

we obtain, for $k_1 \neq k_2$,

$$ye^{k_2 t} = \frac{k_3 x(0)}{k_2 - k_1} e^{(k_2-k_1)t} + C_1,$$

Putting $t = 0$, we have

$$C_1 = y(0) - \frac{k_3 x_0}{k_2 - k_1}.$$

Hence we deduce (5.14).

If $k_1 = k_2$, we have $ye^{k_2 t} = k_3 x(0)t + c_2$, $c_2 = y(0)$, and so we arrive at (5.15).

Example 5.4. An irreversible first-order reaction $A \to B \to C$ in which the reacting substance A changes into the resulting substance C by way of the intermediate substance B is described by the system of DEs

$$\begin{cases} -\frac{\mathrm{d}[A]}{\mathrm{d}t} = k_1[A] \\ -\frac{\mathrm{d}[B]}{\mathrm{d}t} = k_2[B] - k_1[A]. \end{cases} \tag{5.16}$$

We solve this under the initial conditions $[A]|_{t=0} = [A](0)$, $[B]|_{t=0} = 0$ to obtain

$$[A] = e^{-k_1 t}[A](0),$$

and for $k_1 \neq k_2$, $[B] = \frac{k_1}{k_2-k_1}(e^{-k_1 t} - e^{-k_2 t})[A](0)$, while for $k_1 = k_2$, $[B] = [A](0)k_1 t e^{-k_1 t}$.

This is Example 5.3 with $k_3 = k_1$.

Example 5.5. Suppose we have a sequence of radioactive substances whose elements decay successively with distinct decaying constants $\lambda_j > 0$, starting from A_1 and ending with the stable A_n with the number of atoms N_1 and N_n, respectively. The first stage is described in Example 5.3:

$$\begin{cases} -\frac{\mathrm{d}N_1}{\mathrm{d}t} = \lambda_1 N_1 \\ -\frac{\mathrm{d}N_2}{\mathrm{d}t} = \lambda_2 N_2 - \lambda_1 N_1. \end{cases} \tag{5.17}$$

Hence

$$N_2 = N_2(t) = \frac{\lambda_1}{\lambda_2 - \lambda_1}(e^{-\lambda_1 t} - e^{-\lambda_2 t})N_1(0). \tag{5.18}$$

The jth step is described as $-\frac{dN_j}{dt} = \lambda_j N_j - \lambda_{j-1}N_{j-1}$. Hence the second stage is described by the system of DEs

$$\begin{cases} -\frac{dN_2}{dt} = \lambda_2 N_2 - \lambda_1 N_1 \\ -\frac{dN_3}{dt} = \lambda_3 N_3 - \lambda_2 N_2. \end{cases} \tag{5.19}$$

The second equation in (5.19) can be solved by Theorem 1.10 with N_2 as in (5.18) and $a = -\lambda_3$, $u = \lambda_2 N_2$. Hence (1.249) reads

$$\begin{aligned} N_3 = N_3(t) = {} & e^{-\lambda_3 t}N_3(0) \\ & + \frac{\lambda_1\lambda_2}{\lambda_2 - \lambda_1}e^{-\lambda_3 t}N_1(0)\int_0^t (e^{-(\lambda_1-\lambda_3)\tau} - e^{-(\lambda_2-\lambda_3)\tau})\,d\tau. \end{aligned} \tag{5.20}$$

Since $N_3(0) = 0$ and λ_j's are different, it follows that

$$N_3 = N_3(t) = \frac{\lambda_1\lambda_2}{\lambda_2 - \lambda_1}e^{-\lambda_3 t}N_1(0)\left(\frac{1 - e^{-(\lambda_1-\lambda_3)t}}{\lambda_1 - \lambda_3} - \frac{1 - e^{-(\lambda_2-\lambda_3)t}}{\lambda_2 - \lambda_3}\right),$$

or

$$\begin{aligned} N_3 = {} & \lambda_1\lambda_2 N_1(0) \\ & \times\left(\frac{e^{-\lambda_1 t}}{(\lambda_1 - \lambda_2)(\lambda_1 - \lambda_3)} + \frac{e^{-\lambda_2 t}}{(\lambda_2 - \lambda_1)(\lambda_2 - \lambda_3)} + \frac{e^{-\lambda_2 t}}{(\lambda_3 - \lambda_1)(\lambda_3 - \lambda_2)}\right). \end{aligned} \tag{5.21}$$

Similarly, we may deduce explicit formulas for N_j. Now we assume that λ_1 is sufficiently small compared with other λ_j's. Then from (5.18) it follows that $\frac{N_2}{N_1(0)} \sim \frac{\lambda_1}{\lambda_2}$, from (5.21) that $\frac{N_3}{N_1(0)} \sim \frac{\lambda_1}{\lambda_3}$. Hence eventually,

$$\frac{N_j}{N_1(0)} \sim \frac{\lambda_1}{\lambda_j}, \quad 2 \le j \le n. \tag{5.22}$$

Exercise 20. Solve (5.21) following the proof of Theorem 1.10.

Solution. By the second equation in (5.19) and (5.18)

$$\frac{d}{dt}\left(e^{\lambda_3 t}N_3\right) = e^{\lambda_3 t}\lambda_2 N_2 = \frac{\lambda_1\lambda_2}{\lambda_2 - \lambda_1}(e^{-(\lambda_1-\lambda_3)t} - e^{-(\lambda_2-\lambda_3)t})N_1(0). \tag{5.23}$$

Integrating and determining the integral constant to be $\frac{\lambda_1\lambda_2}{(\lambda_2-\lambda_1)(\lambda_3-\lambda_1)}N_1(0)$, we conclude (5.21).

Example 5.6. We take up the successive reactions $^{90}\mathrm{Sr} \to {}^{90}\mathrm{Y} \to {}^{90}\mathrm{Zr}$ in Example 5.2, (iii), (v). The decay constant k_1 for $^{90}\mathrm{Y}$ is $k_1 = 2.826 \times 10^{-6}$

$(\text{hour})^{-1}$. Substituting this and the value $k_2 = \frac{0.693147\cdots}{64} = 1.08 \times 10^{-2}$ $(\text{hour})^{-1}$ of the decay constant in (5.11) gives

$$N_2 = N_2(t) = \varkappa N_1(0) \tag{5.24}$$

where N_2 is the number of atoms of ^{90}Y while N_1 is that of ^{90}Sr and where

$$\varkappa = \frac{2.826 \times 10^{-6}}{1.08 \times 10^{-2} - 2.826 \times 10^{-6}} \times (e^{-2.826\times 10^{-6}t} - e^{-1.08\times 10^{-2}t}). \tag{5.25}$$

Pure ^{90}Sr contains almost no ^{90}Y, but in about a month, it comes to radioactive equilibrium and contains $\frac{1}{3900}$ ^{90}Y, which coincides with the theoretical value

$$\varkappa = 1/3820.65605096 \tag{5.26}$$

computed on the basis of (5.25).

5.3 Other reactions

Definition 5.2. As in Definition 5.1, we refer to an irreversible reaction $A \to B$ as a **$\varkappa$th order reaction**, if the rate of increase $-\frac{d[A]}{dt}$ of B obeys the ($\varkappa$th order) **reaction equation**

$$-\frac{\mathrm{d}[A]}{\mathrm{d}t} = k[A]^{\varkappa}. \tag{5.27}$$

Example 5.7. In a second order reaction suppose that it takes 10 minutes for the concentration of the reactant A to become half of the initial concentration. What will be the concentration after 30 minutes from the start of reaction? The second order reaction is described by the reaction equation

$$-\frac{\mathrm{d}[A]}{\mathrm{d}t} = k[A]^2. \tag{5.28}$$

Solving this, we obtain

$$[A](t) = \frac{1}{kt + [A](0)^{-1}} = \frac{[A](0)}{[A](0)kt + 1}. \tag{5.29}$$

The condition states that

$$\frac{1}{2}[A](0) = [A](10) = \frac{1}{10k + [A](0)^{-1}},$$

whence

$$k = \frac{1}{10}[A](0)^{-1}.$$

Hence (5.29) assumes the form

$$[A](t) = \frac{[A](0)}{\frac{1}{10}t + 1}. \tag{5.30}$$

Hence $[A](30) = \frac{1}{4}[A](0)$. Also half-life T is found from (5.29) as follows.

$$\frac{1}{2}[A](0) = [A](T) = \frac{[A](0)}{[A](0)kT + 1},$$

whence

$$T = \frac{[A](0)}{k} \tag{5.31}$$

i.e. depending on the value of the initial molarity. Note the difference between this and the first-order reaction in Remark 5.1, (i).

Exercise 21. Solve the following reaction equations with initial concentration $[A]_0 = [A](0)$.

(i) The third order reaction

$$-\frac{\mathrm{d}[A]}{\mathrm{d}t} = k[A]^3. \tag{5.32}$$

(ii) The half order reaction

$$-\frac{\mathrm{d}[A]}{\mathrm{d}t} = k[A]^{1/2}. \tag{5.33}$$

5.3.1 *Poisson distribution*

Theorem 5.1. *The functions $x_k(t)$ satisfying the system of differential equations with initial values $x_0(0) = a > 0$, $x_k(0) = 0$, $k \in \mathbb{N}$*

$$\begin{cases} \frac{\mathrm{d}x_0}{\mathrm{d}t} = -\lambda x_0 \\ \frac{\mathrm{d}x_k}{\mathrm{d}t} = \lambda x_k - \lambda x_{k-1}, \end{cases} \tag{5.34}$$

is given by

$$x_k(t) = x_0(0)e^{-\lambda t}\frac{(\lambda t)^k}{k|}. \tag{5.35}$$

Proof. The system (5.34) with $k = 1$ is exactly (5.13) in Example 5.3 with $k_1 = k_2 = k_3 = \lambda$. Hence $x_0(t) = e^{-\lambda t}x_0(0)$ and

$$x_1(t) = x_0(0)e^{-\lambda t}(\lambda t). \tag{5.36}$$

From this inductively, we conclude (5.35). □

Definition 5.3. The probability distribution with the density function

$$Po_k(t) = e^{-\lambda t}\frac{(\lambda t)^k}{k|} \tag{5.37}$$

is called the Poisson distribution.

Example 5.8. Binary distribution with parameters n and p ($n \in \mathbb{N}, 0 < p < 1$) refers to the probability distribution of the number k of occurrences of the specific event with success probability p when we perform n independent experiments. The probability $p(k)$ of the number k of occurrences of the event is given by

$$p(k) = {}_nC_k p^k q^{n-k}, \quad k = 0, 1, \cdots, n,$$

where customarily we write $q = 1 - p$. Since we have

$$\sum_{k=0}^{n} p(k) = \sum_{k=0}^{n} {}_nC_k p^k q^{n-k} = (p+q)^n = 1,$$

$p(k)$ is a probability function.

We say that X obeys the **binary distribution** $B(n,p)$ and write $X \sim B(n,p)$ if

$$P\{X = k\} = p(k) = {}_nC_k p^k q^{n-k}, \quad k = 0, 1, \cdots, n$$

holds.

Example 5.9. In binary distribution with parameters n and p ($n \in \mathbb{N}, 0 < p < 1$ — small), if $n \to \infty$ under the condition that $mp = n$, then we get the **Poisson distribution** $Po(k)$:

$$P\{X = k\} = {}_nC_k p^k q^{n-k} \to Po(k) = e^{-m}\frac{m^k}{k!}, \quad k = 0, 1, \cdots. \tag{5.38}$$

This probability distribution appears when one assembles many events which rarely happen. Examples include the number of occurrences of earthquake, decay of radioactive substances, etc.

The Poisson distribution is the binary distribution with success probability small and number of trials large.

Exercise 22. Check that $Po(k)$ in (5.38) is a distribution function and find the expectation and variance of $Po(k)$.

Solution. $\sum_{k=0}^{\infty} Po(k) = e^{-m}\sum_{k=0}^{\infty} \frac{m^k}{k!} = 1$. By definition $E\{X\} = \sum_{k=0}^{\infty} kPo(k) = e^{-m}\sum_{k=1}^{\infty} \frac{m^k}{(k-1)!}$. Factoring out m, we see that $E\{X\} = m$.

Exercise 23. Find the variance of $Po(k)$.

Solution. By definition

$$V\{X\} = \sum_{k=0}^{\infty} (k-m)^2 Po(k) = e^{-m}\left(-m^2 * m + \sum_{k=1}^{\infty} k\frac{m^k}{(k-1)!}\right).$$

Writing $k = k - 1 + 1$ and dividing the sum into two parts, we find that

$$V\{X\} = m.$$

5.4 The adaptation principle in chemical compounds

5.4.1 *Introduction*

This section is based on the idea of adaptation by which we mean the replacement of one element in a system with another with similar aspects. The first adaptation comes from the replacement of orthophospheric acid H_3PO_4 by arsenic acid H_3AsO_4.

There is a close relation between calcium-sulfate and calcium carbonate-arsenate species. The main objective of this section is to give an explicit formula for the ratio of each arsenal species to the total and to make an adaptation of the former by the latter to develop a new adsorption material of arsenic acid which spends almost no energy. Some recently proposed methods of using calcium sulphate or other material has ambivalent aspects regarding energy spending as well as the treatment after adsorption.

The problem we are facing in the 21st century is decidedly that of getting as much as energy without giving too much burden on the environment. This problem, however, has been unnoticed, or noticed but neglected, by rapidly developing countries which want to attain swift industrial success. This is the track that we have once gone along and we know what will come. Thus, energy and environmental protection are the two sides of a coin and cannot be separated.

In this subsection, as a manifestation of the principle of environmental protection with little energy, we shall consider the adsorption of arsenic acid by calcium carbonate.

Arsenic compounds contained in underground water are highly poisonous and when used as drinking water, it leads to possible damaging of health including blackfoot disease. There are many significant health damages reported in Bangladesh and India, especially in the West Bengal area, as well as the wide area underground water pollution by arsenic compounds in the southern area of Fukuoka Prefecture, Japan, revealed in 1994, most of which are due to natural causes.

The ionic characteristics of arsenic compounds is associated to that of the sodium carbonate and calcium carbonate and higher concentration of arsenic compounds can be seen in deep sodium carbonate type underground water.

Region II shows Calcium carbonate type ($n = 14$, As average 0.023 mg/ℓ) while Region III shows Sodium carbonate type ($n = 14$, As average 0.075 mg/ℓ).

There is a long history of inhibition, by calcium adsorption, of eutrophication of waters, resulting from too much phosphate whose main ingredient is orthophospheric acid H_3PO_4. Kondo [Kondo (1993)] took up the problem of removal of arsenious acid H_3AsO_4 by calcium adsorption. Our study as well as his are based on the **principle of adaptation** to the effect that chemical elements of similar structures and nature may be exchanged, yielding similar results and analysis can be done in much the same way.

In his thesis, Kondo made an extensive experiments on calcium adsorption of arsenic acid on the adaptation principle, i.e. substituting arsenic acid for orthophospheric acid. There were several candidates for adsorption material including coral limestone, activated alumina and calcium carbonate. He used three types of granules of oyster shells all of size smaller than 1 mm: Type I are natural oyster shells, Type II and III are those which are calcined at 600 °C for 2 hours with Type II in pelleted. Experiments were made with As water solution of 20 mℓ with As concentration 50 $\mu g/\ell$. By his thorough investigation, it has been revealed that calcined oyster shells II, III showed 84.6 percent rate of adsorption, almost equal to that of activated alumina. This rate improved for higher concentration of As at concentration 10~2000 $\mu g/\ell$. Type I shells showed 17~37 percent adsorption rate for the use of more than 0.5 g and 2 g the rate was the highest.

The pH value of the resulting water solution was 8.1~9.0 for Type I while it was 9.8~10.7 for Types II and III, surpassing the tap water limit of 5.8~8.6. The third author conjectured that the reason for better adsorption due to the fact that by calcination, calcium carbonate changes into water-soluble calcium oxide CaO (quicklime). He confirmed his conjecture by using the neutralized Type II shells which showed weaker adsorption efficiency. It has been verified that Type II, III shells are very effective in adsorption but have the defect of producing strong alkali solution.

Recently there appeared a paper [Tsunematsu *et al.* (2012)] which proposes the use of calcium sulfate (gypsum) $CaSO_4$ as adsorption material for arsenic acid. This lies at the opposite end of the spectrum of adsorption, yielding strong acidity in addition to sulphur although it has a far stronger in adsorbing power than calcium carbonate.

The same problem occurs with oyster shells which liquate out heavy metal ions contained in them, including Cr, Sr, Mo from calcined ones and Sr, Mo, Mn from Type I, causing a plausible secondary pollution.

Calcined calcium carbonate (quicklime), as adsorption material, suffers from two defects, i.e. it is strong alkali and needs to be burnt at the temperature of $900°\sim1500°$. Shell lime is made by calcination at 300° while the building material is made at 1500°, which gives rise to mortar.

It seems that adsorption of As and liquation of ingredients are in ambivalent balance and we should choose the most economical Type I shells since they are industrial wastes.

Thus Type I shells may be worth further being studied as to why they show the most effective rate for 2 g for water solution 20 mℓ of As concentration 50 $\mu g/\ell$ and why the rate decreases thereafter. This might be caused by the rising of pH of the solution.

There are subtleties of dissociation of which family of acids are prevailing according to the proton index. We restrict to the range 5.8$\sim$8.6 which is allowed as tap water in Japan. Our purpose is to make clear the underlying chemical structure by mathematical reasoning.

This chapter is a starter for extensive research on the adaptation principle. The optimal choice of the adsorption material according to higher concentration of As or its equivalent will be conducted further.

5.5 Adaptation principle

Let X denote one of SO_4^{2-}, HCO_3^-, CO_3^{2-} and F_2^{2-}. Their compound with calcium ion $CaSO_4$, $CaHCO_3$, $CaCO_3$ and CaF_2 are calcium sulfate, calcium hydrocarbonate, calcium carbonate, calcium fluoride, respectively, referred to above. The well-known formula for the **solubility product** reads

$$K_{s,\mathrm{Ca}X} = [\mathrm{Ca}^{2+}][X^{2-}]. \tag{5.39}$$

Specifically, for $X = SO_4$, (5.39) reads

$$K_{s,\mathrm{CaSO_4}} = [\mathrm{Ca}^{2+}][\mathrm{SO_4^{2-}}] = 2.4 \times 10^{-5}\ \mathrm{mol^2dm^{-6}} \tag{5.40}$$

for HCO_3^-,

$$K_{s,\mathrm{CaHCO_3^-}} = [\mathrm{Ca}^{2+}][\mathrm{HCO_3^-}] = 1.02\ \mathrm{mol^3dm^{-9}} \tag{5.41}$$

and for $X = CO_3^{2-}$

$$K_{s,CaCO_3} = [Ca^{2+}][CO_3^{2-}] = 4.8 \times 10^{-9}\ mol^2dm^{-6}. \tag{5.42}$$

We notice the principle to the effect that the smaller the exponent a of 10^a is, the less soluble the material is, i.e. the more precipitative. In the above case, $CaSO_4$ is more soluble than $CaCO_3$.

On the other hand, the solubility product formula for the calcium arsenate $Ca_3(AsO_4)_2$ is

$$K_{s,Ca_3(AsO_4)_2} = [Ca^{2+}]^3[AsO_4^{3-}]^2 = 6.8 \times 10^{-19}\ mol^5dm^{-15} \tag{5.43}$$

By the above principle, calcium arsenate is much less soluble than calcium sulfate and calcium carbonate, i.e. in the presence of the latter, the arsenate ions AsO_4^{3-} precipitate, thereby adsorption can be done. They arise by the dissociation of the protons from arsenic acid in the following successive dissociation equilibrium.

$$\begin{aligned} &H_3AsO_4 \leftrightarrow H^+ + H_2AsO_4^- \\ &K_1 := [H^+][H_2AsO_4^-]/[H_3AsO_4] = 10^{-2.22}\ mol\ dm^{-3}, \end{aligned} \tag{5.44}$$

$$\begin{aligned} &H_2AsO_4^- \leftrightarrow H^+ + HAsO_4^{2-} \\ &K_2 := [H^+][HAsO_4^{2-}]/[H_2AsO_4^-] = 10^{-6.98}\ mol\ dm^{-3}, \end{aligned} \tag{5.45}$$

and

$$\begin{aligned} &HAsO_4^{2-} \leftrightarrow H^+ + AsO_4^{3-} \\ &K_3 := [H^+][AsO_4^{3-}]/[HAsO_4^{2-}] = 10^{-11.4}\ mol\ dm^{-3}. \end{aligned} \tag{5.46}$$

Let As_T denote the total concentration of all arsenic species:

$$As_T = [H_3AsO_4] + [H_2AsO_4^-] + [HAsO_4^{2-}] + [AsO_4^{3-}] \tag{5.47}$$

and let α_j denote the *ratio of each arsenic species against the total*:

$$\alpha_j = [H_{3-j}AsO_4^{j-}]/As_T, \quad 0 \le j \le 3, \tag{5.48}$$

where $\alpha_0 = [H_3AsO_4]/As_T$ and

$$\alpha_3 = [AsO_4^{3-}]/As_T \tag{5.49}$$

which will play the important role. We define similar ratios β_j's and γ_j's below.

Noting that (5.44), (5.45), (5.46) may be unified into

$$K_j = [H^+][H_{3-j}AsO_4^{j-}]/[H_{4-j}AsO_4^{(j-1)-}] = [H^+]\frac{[H_{3-j}AsO_4^{j-}]/As_T}{[H_{4-j}AsO_4^{(j-1)-}]/As_T} \tag{5.50}$$

$$= [H^+]\frac{\alpha_j}{\alpha_{j-1}},\ 1 \le j \le 3,$$

we obtain by (5.50)

$$\alpha_j = K_j[H^+]^{-1}\alpha_{j-1},\ 1 \le j \le 3, \tag{5.51}$$

the relation between successive terms of the sequence, whence inductively we conclude that

$$\alpha_j = K_j \cdots K_1[H^+]^{-j}\alpha_0,\ 1 \le j \le 3. \tag{5.52}$$

Recall that the sum of α_j's is 1 by (5.48)

$$\sum_{j=0}^{3} \alpha_j = 1. \tag{5.53}$$

Substituting from (5.52), we conclude that

$$\alpha_0 = \frac{1}{1 + \sum_{i=1}^{3} K_i \cdots K_1[H^+]^{-i}}. \tag{5.54}$$

Hence from (5.52) and (5.54) we obtain

Theorem 5.2. *We have the explicit formula*

$$\alpha_j = \alpha_j([H^+]) = \frac{K_j \cdots K_1[H^+]^{-j}}{1 + \sum_{i=1}^{3} K_i \cdots K_1[H^+]^{-i}},\ 1 \le j \le 3. \tag{5.55}$$

Or more concretely,

$$\alpha_j = \alpha_j([H^+]) = \frac{K_j \cdots K_1[H^+]^{3-j}}{[H^+]^3 + 10^{-2.22}[H^+]^2 + 10^{-9.2}[H^+] + 10^{-20.6}}. \tag{5.56}$$

Proof. (5.56) follows from the numerical values in (5.44), (5.45), (5.46),

$$K_1K_2 = 10^{-9.2},\ K_1K_2K_3 = 10^{-20.6} \tag{5.57}$$

and by multiplying the numerator and the denominator of (5.55) by $([H]^+)^3$. □

E.g.

$$\alpha_2 = \alpha_2([\mathrm{H}^+]) = \frac{10^{-9.2}[\mathrm{H}^+]}{[\mathrm{H}^+]^3 + 10^{-2.22}[\mathrm{H}^+]^2 + 10^{-9.2}[\mathrm{H}^+] + 10^{-20.6}}$$

$$\alpha_3 = \alpha_3([\mathrm{H}^+]) = \frac{10^{-20.6}}{[\mathrm{H}^+]^3 + 10^{-2.22}[\mathrm{H}^+]^2 + 10^{-9.2}[\mathrm{H}^+] + 10^{-20.6}}, \quad (5.58)$$

Eliminating $[\mathrm{Ca}^{2+}]$ from (5.39) and (5.43), we obtain

$$\frac{K_{s,\mathrm{Ca}X}{}^3}{K_{s,\mathrm{Ca}_3(\mathrm{AsO}_4)_2}} = \frac{[X^{2-}]^3}{[\mathrm{AsO}_4^{3-}]^2}. \quad (5.59)$$

From (5.49), we have

$$[\mathrm{AsO}_4^{3-}] = \alpha_3 \mathrm{As}_T,$$

whence we deduce from (5.59) that

$$[X^{2-}] = R(\alpha_3 \mathrm{As}_T)^{2/3} \quad (5.60)$$

on writing

$$R = R_X = \frac{K_{s,\mathrm{Ca}X}}{(K_{s,\mathrm{Ca}_3(\mathrm{AsO}_4)_2})^{\frac{1}{3}}}. \quad (5.61)$$

In the case of $X = \mathrm{HCO}_3^-$,

$$R_{\mathrm{HCO}_3^-} = \frac{1.06\ \mathrm{mol}^3\mathrm{dm}^{-9}}{(6.8 \times 10^{-19}\ \mathrm{mol}^5\mathrm{dm}^{-15})^{\frac{1}{3}}} = 0.546 \times 10^{-2}\ \mathrm{mol}^{\frac{1}{3}}\mathrm{dm}^{-1}. \quad (5.62)$$

From

$$\mathrm{Ca}_3(\mathrm{AsO}_4)_2 \leftrightarrow 3\mathrm{Ca}^{2+} + 2\mathrm{AsO}_4^{3-}, \quad (5.63)$$

we find that the concentration of As_T is approximately 3/2 of that of Ca_3, whence we deduce the formula

$$[\mathrm{Ca}^{2+}] = [X^{2-}] + \frac{3}{2}\mathrm{As}_T \quad (5.64)$$

as in [Tsunematsu *et al.* (2012), (11)]. Then substituting (5.60), we obtain

$$[\mathrm{Ca}^{2+}] = R_X(\alpha_3 \mathrm{As}_T)^{2/3} + \frac{3}{2}\mathrm{As}_T. \quad (5.65)$$

Substituting (5.60) and (5.65) in (5.39), we deduce that

$$K_{s,\mathrm{Ca}X} = R_X^2 {\alpha_3}^{4/3} {\mathrm{As}_T}^{4/3} + \frac{3}{2} R_X {\alpha_3}^{2/3} {\mathrm{As}_T}^{5/3} \quad (5.66)$$

and we are to solve the quintic equation with coefficients containing the parameter [H$^+$]

$$\frac{3}{2}R_X{\alpha_3}^{2/3}P^5 + R_X^2{\alpha_3}^{4/3}P^4 - K_{s,\mathrm{Ca}X} = 0 \tag{5.67}$$

on writing $P = {\mathrm{As}_T}^{1/3}$. E.g. by the well-known Newton iterative method, the equation (5.67) may be solved numerically.

We are to solve it with $X = \mathrm{HCO}_3^-$, i.e. with the value (5.62):

$$P^5 + \frac{2}{3}R_{\mathrm{CO}_3}{\alpha_3}^{2/3}P^4 - \frac{2}{3}{R_{\mathrm{CO}_3}}^{-1}{\alpha_3}^{-2/3}K_{s,\mathrm{CaCO}_3} = 0 \tag{5.68}$$

with $\alpha_3 = \alpha_3([\mathrm{H}^+]) =$ defined by (5.58). Figure 5.1 shows the graph of $P = {\mathrm{As}_T}^{1/3}$.

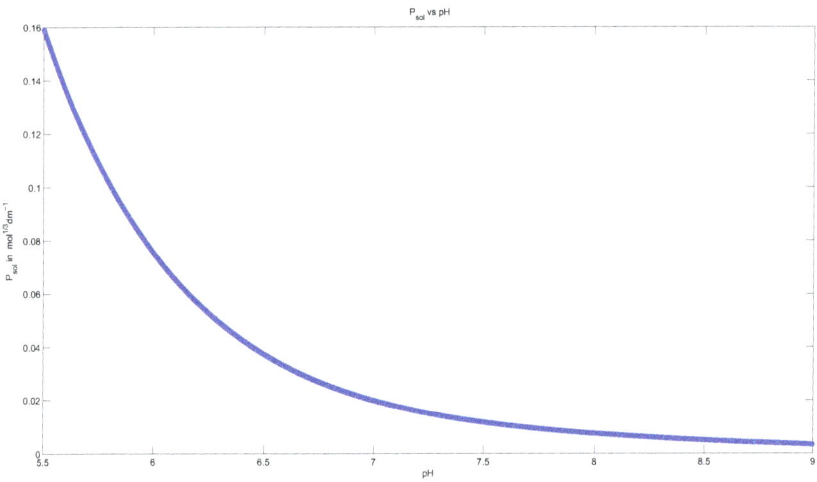

Fig. 5.1: The positive solution of the quintic equation for different values of pH between 5.5 and 8.5

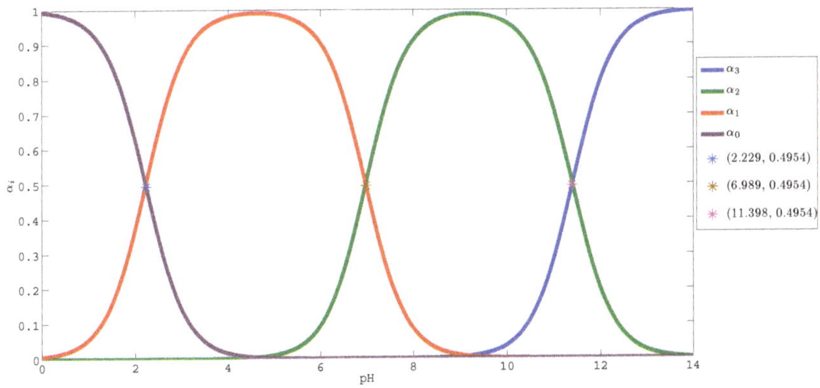

Fig. 5.2: $\alpha_{0,1,2,3}$ value for different values of pH

Remark 5.2. The acidity of tap water is restricted to the range 5.8∼8.6 by law and the underground water has neutral or weak alkali and so the pH lies near and greater than 7. In [Tsunematsu *et al.* (2012)], the pH range is restricted to 7∼10 since $K_2 \approx 10^{-7}$, $\alpha_2 = 10^{-7}[H^+]^{-1}\alpha_1$ and for pH = 7, we have $\alpha_2 = \alpha_1$ and as pH increases, the proportion of $H_2AsO_4^{1-}$ against $HAsO_4^{2-}$ steadily decreases. In [Tsunematsu *et al.* (2012)], the argument goes in the following lines. $HAsO_4^{2-}$ starts predominating $H_2AsO_4^{1-}$, so that one may approximate the amount of As_T by $HAsO_4^{2-}$ and so one may obtain the amount of As_T by solving the equation (5.66) with $X = SO_4$.

In the case of HCO_3^-, it predominates CO_3^{2-} at pH = 8.3 by 100 times and only at pH = 10.3, their amount becomes equal. Hence there are three ways to go. We assume the pH range to 5.8∼8.3, which is most practical or we assume pH > 10.3. Or we calcite the lime stone to produce quicklime. We can make a table of temperature needed for the degree of adsorption.

5.5.1 *Calcium phosphate*

The solubility product formula (5.39) takes on slightly different form according to the compound we consider. For the ease of detection in successive experiments, it has been customary to measure the amount of phosphate rather than arsenal, which should give as good detection result for the latter as the former, in view of the adaptation principle §5.5. In this section we shall state the theoretical values for the ratios (5.69) with P_T indicating the total concentration all phosphate species. Analyis is exactly as in §5.5 but we give some details for readers' convenience.

Theorem 5.3. *Similarly to* (5.48), *let*

$$\beta_j = \frac{[H_{3-j}PO_4^{-j}]}{P_T}. \tag{5.69}$$

Then (5.56) *reads*

$$\beta_j = \beta_j([H^+]) = \frac{K_j \cdots K_1 [H^+]^{3-j}}{[H^+]^3 + 7.5 \times 10^{-3}[H^+]^2 + 4.65 \times 10^{-10}[H^+] + 9.95 \times 10^{-23}}. \tag{5.70}$$

I.e.

$$\beta_1 = \beta_1([H^+]) = \frac{7.5 \times 10^{-3}[H^+]^2}{[H^+]^3 + 7.5 \times 10^{-3}[H^+]^2 + 4.65 \times 10^{-10}[H^+] + 9.95 \times 10^{-23}} \tag{5.71}$$

$$\beta_2 = \beta_2([H^+]) = \frac{4.65 \times 10^{-10}[H^+]}{[H^+]^3 + 7.5 \times 10^{-3}[H^+]^2 + 4.65 \times 10^{-10}[H^+] + 9.95 \times 10^{-23}} \tag{5.72}$$

$$\beta_3 = \beta_3([H^+]) = \frac{9.95 \times 10^{-23}}{[H^+]^3 + 7.5 \times 10^{-3}[H^+]^2 + 4.65 \times 10^{-10}[H^+] + 9.95 \times 10^{-23}}. \tag{5.73}$$

Proof. Correspondingly to (5.63), we have

$$Ca_3(PO_4)_2 \leftrightarrow 3Ca^{2+} + 2PO_4^{3-}, \tag{5.74}$$

so that (5.39) reads

$$K_{s,Ca_3((PO)_4)_2} = [Ca^{2+}]^3[PO_4^{3-}]^2 = 2.0 \times 10^{-29} \text{ mol}^5\text{dm}^{-15} \tag{5.75}$$

(5.44), (5.45), (5.46) read respectively

$$\begin{aligned} &H_3PO_4 \leftrightarrow H^+ + H_2PO_4^- \\ &K_1 := [H^+][H_2PO_4^-]/[H_3PO_4] = 7.5 \times 10^{-3} \text{ mol dm}^{-3}, \end{aligned} \tag{5.76}$$

$$\begin{aligned} &H_2PO_4^- \leftrightarrow H^+ + HPO_4^{2-} \\ &K_2 := [H^+][HPO_4^{2-}]/[H_2PO_4^-] = 6.2 \times 10^{-8} \text{ mol dm}^{-3}, \end{aligned} \tag{5.77}$$

and

$$\mathrm{HPO_4^{2-}} \leftrightarrow \mathrm{H^+} + \mathrm{PO_4^{3-}} \tag{5.78}$$
$$K_3 := [\mathrm{H^+}][\mathrm{PO_4^{3-}}]/[\mathrm{HPO_4^{2-}}] = 2.14 \times 10^{-13}\ \mathrm{mol\ dm^{-3}}.$$

Now (5.57) takes the form

$$K_1K_2 = 4.65 \times 10^{-10},\ K_1K_2K_3 = 9.95 \times 10^{-23}. \tag{5.79}$$

Incorporating these completes the proof. □

(5.65) remains the same with As replaced by PO_4 and we are to solve the quintic equation corresponding to (5.67):

$$\frac{3}{2}R_X\beta_3{}^{2/3}P^5 + R_X^2\beta_3{}^{4/3}P^4 - K_{s,\mathrm{Ca}X} = 0 \tag{5.80}$$

on writing $P = ((\mathrm{PO_4})_T)^{1/3}$.

We are to solve it with $X = \mathrm{CO_3}$, i.e. with the value (5.62):

$$P^5 + \frac{2}{3}R_{\mathrm{CO_3}}\beta_3{}^{2/3}P^4 - \frac{2}{3}R_{\mathrm{CO_3}}{}^{-1}\beta_3{}^{-2/3}K_{s,\mathrm{CaCO_3}} = 0 \tag{5.81}$$

with $\beta_3 = \beta_3([\mathrm{H^+}])$ given by (5.73). Figure 5.3 shows the graph of $P = ((\mathrm{PO_4})_T)^{1/3}$.

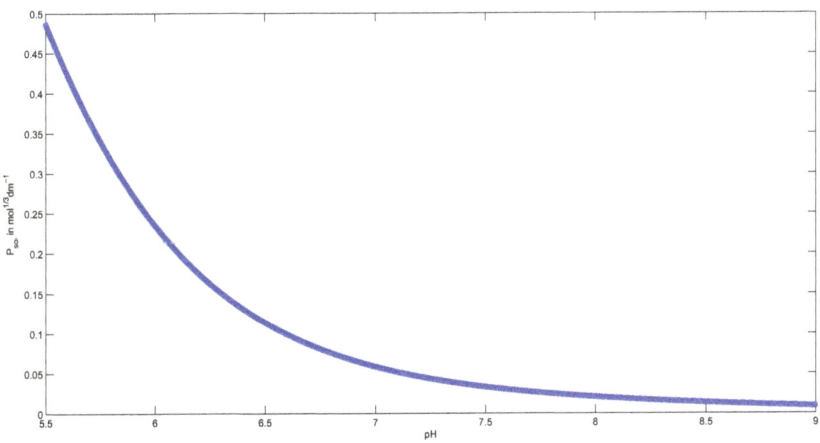

Fig. 5.3: The positive solution of the quintic equation for different values of pH between 5.5 and 9

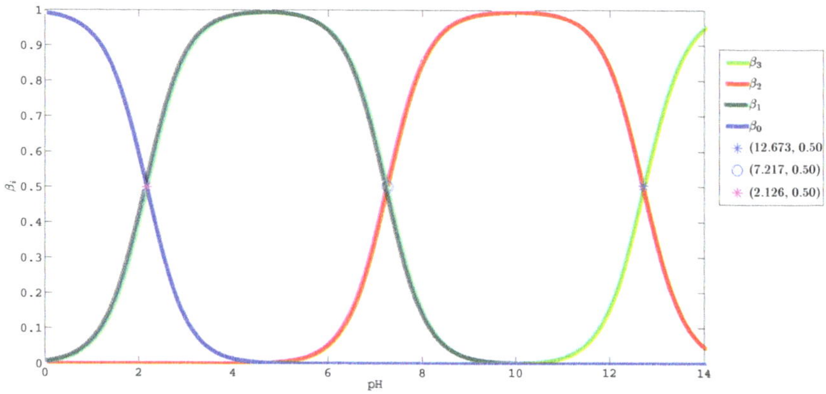

Fig. 5.4: $\beta_{0,1,2,3}$ value for different values of pH

5.5.2 *Dissociation of calcium carbonate*

Since calcium carbonate is weak acid, there is a subtle problem as to which ionic species prevails at a specified range of pH. The following is a preliminary study on the dissociation equilibrium for calcium carbonate. First

$$CO_2 + H_2O \leftrightarrow H_2CO_3, \tag{5.82}$$

which indicates the solution of carbon dioxide into water, and which is irrelevant to us here. What may have to do with are the following.

$$H_2CO_3 \leftrightarrow H^+ + HCO_3^- \quad k_2 = [H^+][HCO_3^-]/[H_2CO_3] = 10^{-6.3} \tag{5.83}$$

$$HCO_3^- \leftrightarrow H^+ + CO_3^{2-} \quad k_3 = [H^+][CO_3^-]/[HCO_3^-] = 10^{-10.3} \tag{5.84}$$

Along with (5.48), we let

$$\gamma_j = [H_{2-j}CO_3^{j-}]/CO_T, \quad 0 \le j \le 2, \tag{5.85}$$

where CO_T indicates the total amount of CO. More concretely,

$$\gamma_0 = [H_2CO_3]/CO_T, \gamma_1 = [HCO_3^-]/CO_T, \gamma_2 = [CO_3^{2-}]/CO_T. \tag{5.86}$$

As with Theorem 5.2, we obtain

Theorem 5.4.

$$\gamma_1 = \frac{10^{-6.3}[H^+]^{-1}}{1 + 10^{-6.3}[H^+]^{-1} + 10^{-16.6}[H^+]^{-2}}, \tag{5.87}$$

$$\gamma_2 = \frac{10^{-16.6}[H^+]^{-2}}{1 + 10^{-6.3}[H^+]^{-1} + 10^{-16.6}[H^+]^{-2}}.$$

Proof. In analogy to (5.50), we sum up to get

$$k_{j+1} = [\mathrm{H}^+]\frac{\gamma_j}{\gamma_{j-1}}, \ 1 \le j \le 2. \tag{5.88}$$

Hence we obtain the relation between consecutive terms

$$\gamma_j = k_{j+1}[\mathrm{H}^+]^{-1}\gamma_{j-1}, \ 1 \le j \le 2 \tag{5.89}$$

whence inductively,

$$\gamma_j = k_{j+1}\cdots k_2[\mathrm{H}^+]^{-j}\gamma_0, \ 1 \le j \le 2, \tag{5.90}$$

$$\gamma_1 = k_2[\mathrm{H}^+]^{-1}\gamma_0, \ \gamma_2 = k_3k_2[\mathrm{H}^+]^{-2}\gamma_0. \tag{5.91}$$

Here we use

$$\gamma_0 + \gamma_1 + \gamma_2 = 1$$

to deduce as we did (5.54),

$$\gamma_0 = \frac{1}{1 + \sum_{i=2}^{3} k_i \cdots k_2[\mathrm{H}^+]^{1-i}} = \frac{1}{1 + k_2[\mathrm{H}^+]^{-1} + k_2k_3[\mathrm{H}^+]^{-2}} \tag{5.92}$$

$$= \frac{1}{1 + 10^{-6.3}[\mathrm{H}^+]^{-1} + 10^{-16.6}[\mathrm{H}^+]^{-2}}.$$

□

In the range 5.8–8.6 of pH which is a regulation by law, both ions exist in notable quantity. E.g. for $pH = 7$, the rate CO_3^{2-} is $\gamma_2 = \frac{10^{-2.6}}{1+10^{0.7}+10^{-2.6}} = \frac{1}{1+10^{3.3}+10^{2.6}} \approx 4.110^{-4}$, which is neglected in [Nishioka *et al.* (2011)]. On p.v of that paper, $10^{-5}{\sim}10^{-2}$ is regarded as negligible quantity.

5.6 How to protect our bodies from toxins

Lorenz (cf. [Fujita (1997), p. 96]) referred to domestication and deterioration of human beings as a cause of the ruin of modern human beings. **Disused atrophy** is universally common in the body functions of living creatures, notably with mammals. Comparing the body function of the wild carnivorous animals and those in a zoo, it is clearly seen that those in the zoos have more feeble functions, e.g. their biting and tearing power is much less than that of wild ones because they are fed suitably cut food (or soft food).

One can also associate the word "Cats' tongue", meaning a cat cannot eat hot food. Indeed, there does not exist hot food in nature. The hottest food would be the prey meat. The word describes the different living conditions of dogs and cats. The former are tied to a stick outside of the house and have no chance to steal hot food being prepared in the kitchen but the latter are rather free to wander around in the kitchen and steal the fresh-out-of-the-oven hot food. And when the cats try the stolen hot food, they burn their tongs, stirring around with complaints.

Domesticated animals usually have inferior body functions than wild ones. Domesticated cats and dogs which lack exercise can suffer from "adult diseases" or more commonly referred to as "diseases arising from the way of life". They get easily tired because of lack of breath or weaker muscles. However this does not mean they live shorter lives than wild ones. Wild cats and dogs cannot live long because of the hardship of getting enough food and the wound that they would get from fighting for their territories.

The same applies to human beings, too. We modern human beings can live much longer than the primitive cavemen or than tribal Papua New Guineans. The Papua New Guineans are reported not to be suffering from pollen allergy and have very strong and developed muscles in spite of their poor food with less proteins than the civilized men. The reason has been found out by Professor K. Fujita [Fujita (1997)] to be their parasites. Their immune system is too busy with fighting against the parasites to create anti-body against less harmful attacks by pollens or the similar. And these parasites somehow transforms their vegetarian food into proteins, like herbivorous animals.

Body Mass Index (BMI) is defined to be the ratio of the body weight to the square of the height: BMI $= \frac{W}{t^2}$.
An example: Mr. Maeda's BMI $= 19.33$; $23\sim24$ = WHO standard BMI; over 30 is thought of as fat; under 19 thin (slim); Japan standard BMI $=$ 22; over 25 fat; under 18.5 thin.

BMI can be used to check the health condition–fatness in place of the standard body weight.

One of the best ways to keep good health is to take good weak-alkaline water, which is not easy to find (cf. [Hayakawa (2000)]). The pH of our bodies is kept weak alkaline $7.35\sim7.45$. Therefore it is essential to keep our blood weak alkaline–**alkalosis**– to keep good health and sustain our lives. The blood becomes the state of **acidosis**–acidic–when we are exhausted or take acidic food including meat and sugar. It is known that if the blood is in acidosis, then the calcium in bones and teeth start melting in the blood. If calcium adheres to the blood vessel, it hardens the vessel wall and can be the cause of heart attack or brain tumor. Equally important is the low degree of **oxidation-deoxidation potential**, which we refer to as **reduced** water or ODP for brevity. If ODP is higher, then it weakens the functions of enzymes and anti-oxidation substances in the body, so that it creates more **active oxygen** (sometimes activated oxygen, reactive oxygen) which is the main cause of cancer and aging.

We cannot avoid being radiated after the Fukushima, the radioactive pollution going on.

5.6.1 *Monounsaturated fats and active oxygen*

We have always been exposed to toxins including pesticides and environmental pollution. After the Fukushima Daichi nuclear disaster, one more factor is added, i.e. that of the radiation. As is presented in the 3–4 typical radiative substances–Cesium 137, Strontium 90, Plutonium (in the beginning there was also Iodine which are much less now because of its short half-life). Half-lives of Cesium 137, Strontium 90 are about 30 years and even in 30 years, there still remains half of the amount. In the case of Plutonium, the half-life is 25,000 years and it will continue to release radiation.

Strontium 90 is absorbed in bones because **it resembles Calcium** Ca while cesium 137 is absorbed in muscles because **it resembles Potassium** K.

A possible misuse of TFA as Cis-FA in membranes is exactly the same as the case with radioactive substances. The human body constructs cell membranes from Cis-FAs and also uses them as metabolic pathways for hormones. There are 6$\sim$7 trillion cells in the human body.

The plausible danger of TFA has been recognized and there is a purging campaign of TFA in the US. TFA are unnatural FA that do not exist in nature and it is not known how the human bodies use them. If the body use them in the same way as it uses cis-FA, then it may result in damaging cell membranes and hormones. Here a study is necessary of cis-configuration, skew, and trans-configuration. They may coexist chemically. Here arises the problem of similarity: Human bodies can absorb those toxic (mainly artificial) materials which resemble natural ones.

5.6.2 *Cis vs. Trans FA*

We are interested in the difference between saturated and unsaturated fatty acids and what happens when they are oxidized.

This chapter is a starter for research and practice of wholistic medication toward longevity in good health. We shall confine ourselves to the lipids and active oxygen. We are motivated by [Weil (1995)] and others to improve the quality of meals by the use of monounsaturated fat or monounsaturated fatty acid. Hereafter we often abbreviate fatty acids by **FA**'s.

The dictionary [Longman (1978)] contains the following description of "saturated fat".

A kind of fat from meat and milk products that is thought to be less healthy than other kinds of fat from vegetables or fish.

On [Weil (1995), p. 223] we find passages. **The human body constructs cell membranes from cis-FAs and also uses them as metabolic pathways for hormones.** Unsaturated FA are feasible to undergo transformation from natural cis-configuration into unnatural trans-configuration when they are processed by heating, chemical solvent or perching, thus creating TFA—trans-fatty acid. cis is a Latin word meaning 'this side', and trans means 'that side'. The plausible danger of TFA has been recognized and there is a purging campaign of TFA in the US. TFA are unnatural FA that do not exist in nature and it is not known how the human bodies use them. If the body use them in the same way as it uses cis-FA, then it may result in damaging cell membranes and hormones since there are 6$\sim$7 thousand billion $= 10^{12}$ cells in the human body whose membranes play essential roles in maintaining life.

Here a study is necessary for cis-, skew, and trans-configuration. They may coexist chemically.

5.6.3 *Classification of lipids*

Three main ingredients of human bodies are proteins, sugar and lipids.

Lipids are composed of fatty acids which are complexes of carbon, hydrogen and oxygen atoms and are classified into three classes according to the existence of double bonding (of carbon atoms). Those without double bonding are called **saturated fatty acids** and those with double bonding are called **unsaturated fatty acids**, which are then classified into two subclasses depending on the number of double bonding. Those with one are called **monounsaturated fatty acids** and those with more than two are called **polyunsaturated fatty acids**.

Saturated FA's exist in plenty in animal meat including beef, pork, lamb, (unskinned) chicken, duck, etc. and they have the property that they are in solid in room temperature. This comes from the fact that those livestocks (poultry) have *higher body temperature than humans* and their fat gets solidified at lower temperature. We must keep in mind that *animal fat has a tendency to get solidified and therefore may get stuck in blood vessels.*

It seems that unsaturated FA is not well recognized by the public. E.g. [Supp (2008), Fig. 15, p. 33] there is a passage. Among unsaturated FA, those with one double bond are called monounsaturated FA and those with more than two double bonds are called polyunsaturated FA. *These polyunsaturated FA are further classified according to the position(s) of the double bond. The oleic acids which are plenty in olive oil etc. have the double bond at the 9th position from the methyl-base at the left end in its structure formula and are called $n9$-type. Similarly, α-linoleic acids ($n3$-type) and linolenic acids ($n6$-type) have the double bond at the 3rd and the 6th positions from the left, respectively.*

Apparently here is a serious mixing-up occurring of monounsaturated FA and polyunsaturated FA which are to be severely separated as in Fig. 16 [Supp (2008)].

"The oleic acids which are plenty in olive oil etc. have the double bond at the 9th position from the methyl-base at the left end in its structure formula and are called $n9$-type" is true if it's not mixed together since it is monounsaturated FA. But the subsequent passage is wrong in view of the fact that "α-linoleic acids ($n3$-type) consist of 18 carbon atoms with 2 double bonds" and "linolenic acids ($n6$-type) consist of 18 carbon atoms with 3 double bonds". Therefore, in spite of their names, $n3$-type has not only one double bond at the 3rd position but also the other double bond and $n6$-type has not only one double bond at the 6th position but also two more double bonds.

Neither α-linoleic acids nor linolenic acids are synthesized in the human body and should be taken in through food; thus they are **indispensable FAs**. What is important is their ratio, the ideal ratio being 4 : 1. However, modernization and Westernization of meals makes the ratio quite bad at 14 : 1.

From these indispensable FAs, the body synthesizes **prostaglandin** (PG) which is a physiology-activating substance and is responsible for regulating inflammation and immune system. There are 3 kinds of PGs, which are further classified into subclasses.

Linolenic acids transforms into γ-linolenic acids, which further transforms into dihomo-linolenic acids, from which $PG1$ are produced. Dihomo-linolenic acids transform into arachidonic acids from which $PG2$ are synthesized, among which there is $PG\,E2$ which is said to weaken the function of the immune system. Since arachidonic acids are contained in animal meat (beef, pork, chicken), it is advisable to refrain from taking too much animal meat.

On the other hand, α-linoleic acids ($n3$-type) transform into IPA (EPA) which coexist with DHA. Here IPA is Icosapentaenoic acids with EPA Eicosapentaenoic, while DHA is Docosahexaenoic acids. Both of them are contained in fish meat. IPA is in sardines, mackerels and DHA in saurels (horse mackerel)s, bonitos, albacore tunas.

However, it is emphasized in [Weil (1995), pp. 178-179] that Omega-3 FAs should be taken, where Omega-3 and $n3$ are synonyms. It is asserted that the oily cold sea fish—sardines, herring, mackerel, bluefish, salmon, and lesser extent albacore tunas.

Japan is blessed with a rich supply of these fish. For Japanese tables, the traditional fish including **sardines, mackerels, saurels, salmon** are good and cheap supplies of Omega-3 FAs.

Omega-3 FAs are also highly contained in plant fats—flax and hemp. **Flax** oil is becoming famous as a rich supply of Omega-3.

Finally, although it may become persistent weed in your garden, **purslane** could be used as another supply of Omega-3. In Mediterranean meals, this is often used.

On [Weil (1995), p. 174], mention is made on the degeneration of unsaturated FAs into TFA referred to in §5.6.2 under **heating or treating with chemical resolvents and bleaches**. From our simple common sense, it can be perceived that the less the double bond in unsaturated FAs is, the less the damage to them. Thus it seems true that monounsaturated FAs are most resistant to heating and other treatment. The main resources of supply these FAs are olive oil, canola oil, peanut oil and avocado oil. Canola derives from the word Canada (in India) and canola oil is extracted from rape seeds. Canola oil which are sold at low prices are all extracted by deforming the FAs and also rape seeds may be contaminated by pesticides. Only the costly one which is **organic expeller-pressed** canola oil is recommendable. Olive oil is far safer as there is no need of pesticide and has been used safely for thousands of years.

Olive oil of good quality—**extra virgin olive oil**—has become available easily at reasonable price and it is highly recommendable to use it for frying in deep oil. Especially, in Japanese cooking, frying food material in deep oil is very common and if you use olive oil for it, you'll be freed from sour stomach after eating.

$$\text{FAs}\begin{cases}\text{saturated FA}\\ \text{unsaturated FA}\begin{cases}\text{monounsaturated FA}\\ \text{polyunsaturated FA}\begin{cases}n6\text{ type}\to\gamma\text{linoleic acid}\\ n3\text{ type}\to\text{IPA}\end{cases}\end{cases}\end{cases} \tag{5.93}$$

monounsaturated FA

oleic acids (n9-type): $C_{17}H_{33}COOH$

polyunsaturated FA

linolenic acids (n3-type): $CH_3(CH_2CH = CH)_3(CH_2)_7COOH$

linoleic acids (n6-type): $CH_3(CH_2)_3(CH_2CH = CH)_2(CH_2)_7COOH$

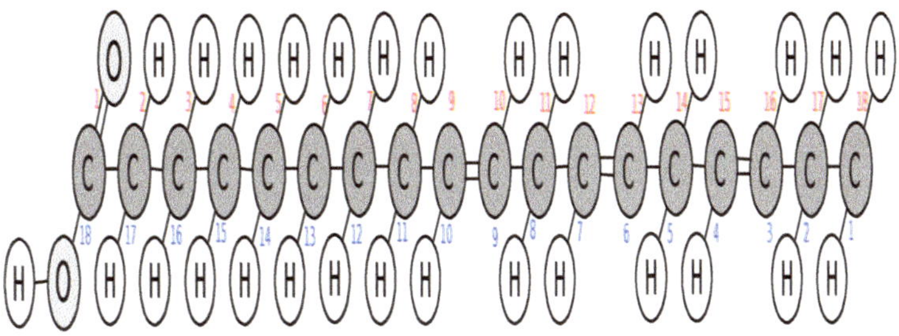

Fig. 5.5: ω-3,6,9 FA

5.6.4 *Comparison of FA's*

Theretofore, however, saturated fats and polyunsaturated fats have been mainly used. Animal fats are highly saturated and the higher the saturation degree, the higher the temperature at which the fats melt. The main natural resources of saturated FAs are products made from whole milk including typically butter, cheese and cream and processed foods made with two highly saturated vegetable tropical oils (coconut, palm) as well as animal meat, as mentioned in §5.6.3, and whole milk themselves.

Saturated FA's are in solid state in room temperature. In the case of animal meat, it is partly because those livestocks (poultry) have **higher body temperature than humans**, sometimes reaching over 40° and their fat gets solidified at lower room temperature of 28°. See §5.6.8 for more elucidation.

Hence *animal fat has tendency to get solidified and therefore may get stuck in blood vessels*, which would rather clearly explain the harm of too much taking of animal fat resulting in cardiovascular diseases.

Indeed, more theoretical explanation of solidity of saturated FA's is that they have no double or triple bonding of carbon atoms and so they are more stable than unsaturated ones.

Lying in the opposite side of saturated ones in the chemical spectrum of fats are polyunsaturated fats, which are in liquid state at room temperature. Corn, soy, sesame, sunflower and safflower oils are typical examples of polyunsaturated FA's. They are all of vegetable origin and in view of our imprinted common sense, we tend to believe that vegetable fats are healthier than animal fats. Thus once margarine drove off butter from the throne of pasting oil on bread on this ground.

However, margarine is a saturated FA by partially hydrogenating the unsaturated oils. As is discussed in §5.6.3, when processed chemically, unsaturated FAs become TFA which are much more harmful than saturated FAs. Thus margarine is more harmful to heart disease or some degenerative changes to the body than butter. A strategy is to remove margarine and solid vegetable shortening and products made from them from the diet. "Crisp" feeling of cookies or biscuits is the result of shortening. Thus homemade cookies are healthier.

5.6.5 *Molecules of the shape HAAH*

The following is a schematic expression of the shapes of molecules of the form HAAH, cf. [Gimarc (1979)], [Yoshida (1956), pp. 73-88].

acetylene C_2H_2 C-H≡C-H with 10 valence electrons is the most stable in linear shape.

diimine N_2H_2 (H-N=N-H) with 12 valence electrons is the most stable in planar cis- or trans-shape.

hydrogen peroxide H_2O_2 (H-O-O-H) with 14 valence electrons is the most stable in skew shape.

molecule	valence electrons	structure
acetylene C_2H_2	10	linear $D_{\infty h}$
diimine N_2H_2	12	planar cis- or trans- C_{2v} or C_{2h}
hydrogen peroxide H_2O_2	10	skew C_2

Table 5.1: HAAH molecules

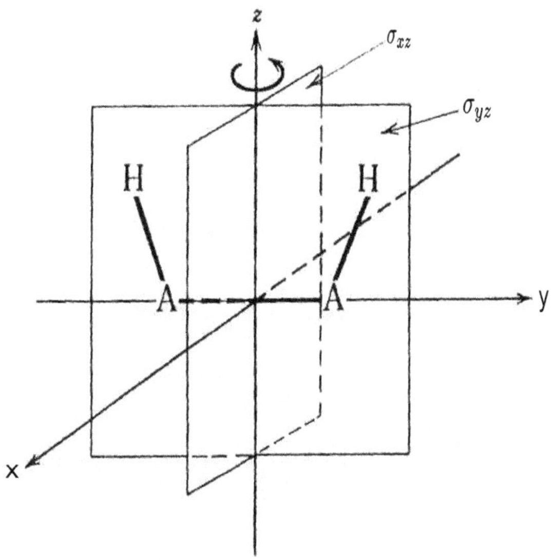

Fig. 5.6: Cis-configuration

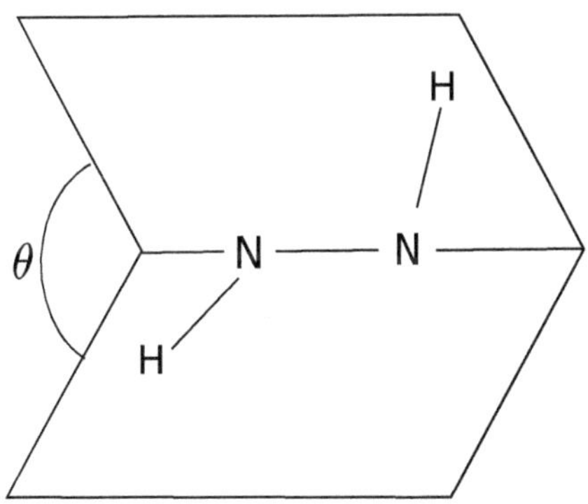

Fig. 5.7: Skew-configuration

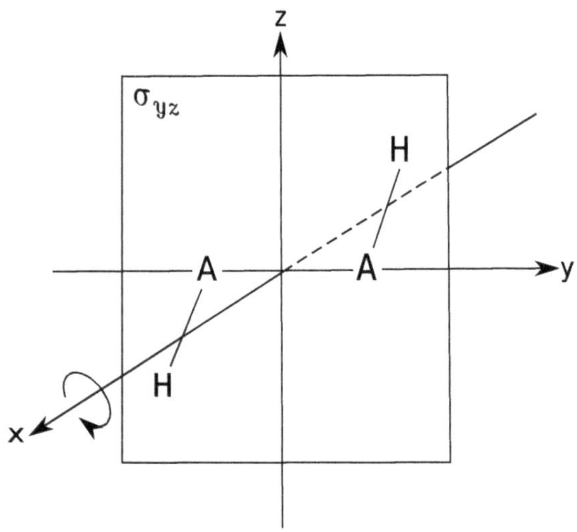

Fig. 5.8: Trans-configuration

5.6.6 *Strength of intermolecular forces*

On [Carbone and Gromov (2001), p. 17] etc. one finds the following table.

- covalent $C = C$ bond 20000
- covalent phosphate bond between nucleotides 5000
- ion-ion interactions 1000
- hydrogen bonds 100
- dipole-dipole interactions 10
- van der Waals interactions 1,

the numbers indicating the approximate ratios. Energy of hydrogen bond is approximately 1 kcal/mol. The table shows that covalent carbon-carbon covalent bond is very strong, explaining the reason why the saturated FA's are stable and keep solid state at low temperature of 28 °C.

5.6.7 *Rezension von Bibliographie*

One cannot but get surprised at so many fallacious statements and one-sided views included in so many references, some of which we mentioned in the above. Here we assemble others which are of less importance but could be harmful as demagogy to the public.

(1) In TV programs, one most often sees many statements which are true in only on one side. E.g. eating cold cooked rice may not add more weight because the rice is harder to digest. To keep the weight, it may sound good but there is a more serious factor hidden here. Being poorly digested means that it gives more burden on the digestive system and eventually on homeostasis. If one keeps such kind of diet in youth, the result in the long run would be rather fatal.

(2) In [Supp (2008), pp. 41-42] we find the four classes of AOs in the order. Superoxide(-radical) $<$ hydrogen peroxide (H_2O_2) $<$ Hydroxylradical $<$ singlet oxygen, where the inequalities mean the strength of oxidation. We shall debunk the fallacy in this by claiming that this is a wrong order and their damage on living organisms is far more complicated.

5.6.8 *Debunking fallacies by persuasive reasoning*

In order to enlighten the public, it may be necessary to use rather persuasive phrases. We will collect some of them here.

(1) As we have referred to in §5.6.4,

"Saturated FAs existing in animal meat including beef, pork, (un-skinned) chicken etc. are in solid in room temperature. The reason is that those livestocks have higher body temperature than humans and their fat gets solidified at lower temperature."

is so persuasive that people would understand such FAs would give trouble to their blood vessels.

(2) There are 6~7 thousand billion $= 10^{12}$ cells in the human body and cell membranes play essential roles in maintaining life.

(3) On [Kawashima (1982), pp. 26-27] we find a question whether meat and pickled plum are acidic or alkali food. A first answer would be meat is alkali and pickled plum is acidic from their taste. The answer is proved to be wrong by heating them. If we keep heating them, there will remain white ashes in which there are contained phosphate and sulfur in the case of meat and Calcium, Potassium, Sodium (Natrium) and Magnesium. The former elements turn into acid, phosphoric acid and sulfuric acid while the latter minerals are alkali. Therefore, meat is acidic and pickled plum is alkali! The sour taste of pickled plum is caused by organic acid such as citric acid, which vapors out when heated. In our bodies the same process occurs at a slower pace but what remains is the same and if we eat meat, then our blood turns into acidic—acidosis.

5.7 Active oxygen species—AOs

Toxicity of oxygen to living organisms are mainly due to its reductive products **superoxide** O_2^-, **peroxide** O_2^{2-}, **hydroxyl radical** ^1HO etc., see below. The peroxide is *less acidic* than the superoxide, and exists in the neutral acidity range as the conjugate acid hydrogen peroxide. The superoxide has pK_a 4.8, which is almost the same as **acetic acid** $HOOCCH_3$ (CH_3COO^-) and exists in the neutral acidity range as O_2^-.

AOs oxidize lipids to make them **lipoperoxide** (peroxidized lipids) As on [Ikeda (1993), p. 97] the resultant lipoperoxide may attach to the inner membranes of articular vessels to be a cause for worsening of and also as one of main causes for liver inflammation.

We must notice the principle of **cancer therapy by radiation**. The hydroxyl radical $^\bullet HO$ is produced not only from peroxide but also when radiation acts on water. The cancer therapy depends on this principle that *hydroxyl radical damages the genes of cancer cells to make them die.*

The danger of radiation is that when our bodies are exposed to it, hydroxyl radical $^\bullet HO$ *is produced which damages the genes.*

As is mentioned in Preface, we are all descendants of those small group of 20,000 members, Mitochondria Eve, who lived 143,000 years ago. All other species extinguished. Since the genes are the career of perpetual information, it is very probable that their genes got damaged and could not reproduce their species. If our genes are kept exposed to radiation, it is highly probable that our species will also extinguish.

These reductive products from oxygen is highly reactive and oxidize those cells existing nearby one after another. They become stronger as they absorb electrons. We refer to the active oxygen species as **AO**'s hereafter. The AO's have different oxidation strength.

superoxide O_2^- < **peroxide** O_2^{2-} or **hydrogen peroxide** H_2O_2 < **hydroxyl radical** ^1HO.

The hydroxyl radical is 100 *times stronger than superoxide.* They arise according to the following chain of reactions.

$$O_2^- \to H_2O_2 \to {}^\bullet HO$$

The **singlet oxygen** $^1O_2$ is the strongest but it arises when exposed to UV. It is one of the main sources for skin aging. (Most of oxygen is triplet oxygen $^3O_2$.)

On [Supp (2008), p. 41] it is stated *wrongly* that AOs are originally supportive killers which white blood cells produce in order to protect the body from the invading bacteria and viruses. On [Sena and Ohta (2007)] there is a remark that another important source of AOs is mitochondria which fact is not well-known. It is inevitable for mitochondria to leak electrons, being a high voltage electric plant.

5.7.1 *Annihilators of AOs*

There are natural detoxificators, called **scavengers**, of activated oxygen in the body. **Superoxide dismutase** SOD annihilates superoxide O_2^-. **Catalase, Glutathione peroxidase, Cytochrome c peroxide enzyme** annihilate hydrogen peroxide H_2O_2.

There is no natural annihilator of hydroxyl radical $^\bullet HO$. Indeed, in radiative treatment of cancer the therapy consists in creating hydroxyl radicals in the body by beaming radiation. They cause apoptosis of the normal cells as well as cancer cells. This is why one feels unwell with radiative treatment.

As one gets older, one has less scavengers and AO start to show their bad effect on the body function.

5.8 Homeostasis

Homeostasis may be defined to be chemo-dynamical stability of cells (whence body functions) in a variable environment, [Carbone and Gromov (2001), p. 44]. Homeostasis w.r.t. the amount of AOs is preserved by the balance of AOs and their scavengers.

Also we are to note the following. The balance between the amount of NO or $NO^\bullet$ and superoxide O_2^- controls the blood pressure. Thus for people with higher blood pressure may absorb NO to reduce the blood pressure: *Say yes to* NO *program*, [Ignalo (2007)].

Thus to keep homeostasis we need a good balance of two ambivalent entities.

As a "hydrogen highway", the effect of hydrogen molecules were reported as strong detoxificators of hydroxyl radical. Hydrogen molecules detoxify the hydroxyl radical in the following process.

$$\begin{aligned} &Fe^{2+} + H_2O_2 \to {}^\bullet OH + Fe^{3+} + OH^- \\ &{}^\bullet OH + H_2 \to H_3O^\bullet \\ &H_3O^\bullet + {}^\bullet OH \to H_4O_2 \to 2H_2O. \end{aligned} \tag{5.94}$$

However, we are to note the following discrimination of statements. On [Sena and Ohta (2007), pp. 244-245] it is stated that the detoxification of hydrogen peroxide in mitochondria is essential in extending the life span (of mice). On the other hand, the hydrogen highway is for detoxifying hydroxyl radicals.

Similarly, the energy production is done by *mutually conflicting processes, anaerobic* **glycolytic pathway** *and the successive aerobic* **oxidative phosphorylation**. Both produce energy in the form of ATP's **Adenosine Triphosphate** (Adenosine 5′-Triphosphate) but the latter produces 36 ATPs while in the former only two ATPs produced. ATP is thought of as an energy currency in living organisms. For when the third phosphate gets liberated, 7.3 cal energy is released from ATP 340 g, and ATP becomes ADP **Adenosine 5′-Diphosphate**.

When glucose (starch) is absorbed in cells, it is transformed into two pyruvic acids and two APTs in cytoplasm sol, which is done in anaerobic conditions. If no oxygen is provided, pyruvic acids are metabolized into lactic acids in the case of humans while ethanol is produced in the case of yeast.

If enough oxygen is supplied at this stage, then pyruvic acids are metabolized in mitochondria into acetyl CoA which is then metabolized through citric acid systems and respiratory chains into CO_2 and water. In the process, ATPs are produced.

A prototype of this phenomenon in the case of beer yeast was first observed by L. Pasteur and so it is called the **Pasteur effect**.

5.9 Circulation of nitrogen

In Chapter 6, we described circulation of carbon. In analogy to this, we consider the circulation of nitrogen, which is plenty, occupying more than 3/4 of the atmospheric gases (about 78%).

We discuss two examples in which nitrogen plays an important role as one of triplets. One is the three essential ingredients in fertilizers, **Nitrogen, Phosphoric acid, and Potassium** while the other is the three essential ingredients of food, **Carbohydrate, Protein, and Lipids**.

5.9.1 *Circulation of nitrogen as fertilizers*

[Seymour and Girardet (1988), p. 48] describes how bacteria fix nitrogen. Nitrogen exists in the living organisms as protein ingredients or tissues. Plants cannot use it directly and they need it in the form of **nitrate** - ONO_2, i.e. nitrogen combined with or **fixed by** oxygen. The only organisms that can do this fixing nitrogen are bacteria, which exist freely in soil or are housed in the root nodules of leguminous plants (pea family including beans, alfalfa, clover, lucerne, lupins). Therefore to breed nitrogen-fixing bacteria in the soil, it is efficient to plow green manure made of those leguminous plants in it in spring.

In the well-tilled soil, the nitrogen-fixing bacteria which either live freely or in the colonies in the root nodules of leguminous plants provide a constant supply of nitrogen which plants can use. However, for agri-business, this process is too slow and so misguided farmers appeal to the artificially fixed nitrate in the form of commercial fertilizers. However, only limited amount of nitrogen is used by plants and the excessive use of nitrogen fertilizers result in abundance of nitrate, which, combined with hydrogen, produces acid soil.

The **solubility equilibrium** of **gypsum plaster** is given by

$$CaSO_4 \rightleftharpoons Ca^{2+} + SO_4^{2-}. \tag{5.95}$$

Similarly to the nitrate case, this produces acidity in the soil. Then there is a need of alkaline fertilizers to neutralize it.

The **ammonium sulphate** $(NH_4)_2SO_4$ has been extremely used as typical artificial fertilizer of nitrate. However, the sulfur that remains after the plants use nitrogen, form gypsum sulphate as in the reaction (5.95) and it hardens the soil which is of no use for agriculture.

5.9.2 *Circulation of nitrogen as food ingredient*

The importance of use of proper lipids in the human bodies has been developed in [Sakurai (2011)] and here we discuss the importance of nitrogen as food ingredients. It is contained in proteins which is a long chain of amino-acids. Although there are indispensable amino acids which must be taken in as food, excessive ingestion of protein results in weakened function of immune system. This follows from the decomposing reaction of amino acids in the body. Protein molecules are big and complicated, and digesting them gives a burden on digestive organs. If one takes more proteins than necessary (60 g per day), then the digestive system must do more work than in the case of other nutrients, carbohydrate and lipids. In the process of metabolism, protein degrades to **highly toxic nitrogenous residues**. The liver being responsible for detoxifying taken-in material, it has to deal with these nitrogenous residues into **urea** $CO(NH_2)_2$ which is a simple compound and is also highly toxic to the body. Then kidneys transform them into **uric acid** $C_5H_4N_4O_3$ which is excreted as urine through urethrae. However, these organs also work as strong allies of the immune system, so that if they are too much engaged with treating proteins, then their function as immune allies will be limited. Therefore living on proteins is not a healthy way at all.

Nitrogenous-breakdown products of protein metabolism irritate the immune system, increasing the risk of allergy and auto-immunity.

Cf. [Weil (1995), pp. 180-182].

Chapter 6

Climate change and carbon circulation

Abstract

We are living at the bottom of the deep sea of air surrounding the Earth. It consists of a few layers of 100 km which is extremely thin compared with the diameter of earth of 12,768 km. The nearest to the surface of the earth is the **troposphere** which ranges from 7 km (near the poles) to 18 km (near the equator). It is said that 90% of climate changes occur in this troposphere. The troposphere is bounded by stratopause above which there is **stratosphere** up to 50 km. It is bounded by mesopause above which there is the mesosphere. In the stratosphere, temperature goes up with the height since the ozone layer absorbs UV. The ozone depletion problem is that the destruction of ozone layer causes and allowing more UV reaching the surface of the Earth. **Global warming** is, by definition, the phenomenon that the temperature of atmosphere, in particular, of the near-surface air, increases.

It has been claimed loudly that the increase of carbon dioxide will cause global warming and one should try to save energy. But at individual or family level, this kind of effort turns out to be ineffective since the main source of greenhouse gases are factories and traffic.

The mass communication also agitates the problem as soon as there is any unusual climate occurs somewhere in the world, e.g. unusually hot summer in Europe, more frequent occurrence of localized downpour, tornado. This type of excessive reaction to minor disturbances is typical of human nature which always wants to find an object of accusation and has no proper attitude of the scientific mind. The global warming problem is a manifestation of the **conflict between energy generation and environmental protection**, the main issue in the 21st century.

We want to know the state-of-the-art affairs of the global warming problem confronting us from the standpoint of Paleoclimatology and will eventually view climate change as a mode jumping into a dynamical system.

We appeal to the most recent discoveries in Paleoclimatology (archeo-climatology). This made a big progress thanks to the ice-core from Greenland. Without development of the boring technology such a core could have never been cut out. To our first-order surprise, the ice-core contained the history of temperature on the earth in the last 100,000 years. Before such a decisive data, the subject used to depend on the record kept in annual rings of old trees—**dendro-climatology**, strata, or fossils. Similarly to the measurement of isotopes of carbon in our present main research interest—carbon circulation— in the study of change of temperature, the measurement of the stable isotope of oxygen plays a central role. This depends on the principle that the higher the temperature, the more heavier oxygen appearing.

From such geo-data, one may conclude a right prediction of the future. The conclusion is that the main issue of the global warming is the speed of increase of carbons in the air, which is **1000 times quicker** than in the past as is discussed in §6.3. Indeed, in the Cretaceous period: $146,000,000 \sim 65,000,000$ years ago, the climate was far warmer than now and the density of CO_2 was 10 times as much as the present one. The change of carbon dioxide by nature is 100 ppm at the time scale of tens of thousands of years such as between the glacial-interglacial period. But in the last 300 years after the Industrial Revolution, the same amount of carbon dioxide were increased by the human activity.

6.1 Prometheus II Era

We concentrate on the energy generation following partly [Murota (1984)]. It depends mostly on fossil fuels and by burning them we produce energy. In this way, we are spending the legacy of the Earth which it assembled over millions of millions years. In the pre-dawn of human beings it was the fire that was kept in the caves of our ancestors. This type of heat source is called **Prometheus I**, after the Greek myth of the giant Prometheus who stole the fire from the gods to give to humans. The feature is that when we burn fuels, the combustion goes on and we get large amount of calories, however, the amount depending on that of fuels used.

In the 18th century, coal started being used instead of firewood. If one just uses coal for getting heat, then it remains in the Prometheus I framework. However, the invention of steam engines made it possible to use small amounts of coal to make them work to get more coal, thus making "coal re-creates more coal." This type of heat source is called **Prometheus II**. The feature is that **fuel that is spent re-creates itself in larger quantity**. Coal has bigger entropy than oil, and the latter has been used for more than 100 years. Both coal and oil are of Prometheus II type heat sources. They are used in **thermal power station** and at home.

Then nuclear power using uranium started being used as supplements to coal and oil, by the reasoning that it is a clear energy source. This type of propaganda is totally false and nuclear power is of utmost danger as has been demonstrated by the Chernobyl [Pohl (1987)], Fukushima [Sakurai (2011)], etc. In addition to its intrinsic menace, nuclear power remains Prometheus II type because it re-creates itself in larger quantity in the form of Plutonium, which is far more dangerous than uranium. Part of the reasons of still sticking to nuclear power is that those developed countries which possess nuclear power plants want to keep large stocks of Plutonium for apparent reasons.

Electricity produced by nuclear reactor is a very large amount and nuclear power plants need back-up electricity sources. As with thermal power plants using coal, nuclear power plants are **stiff** type because once started in generating electricity, it cannot be stopped immediately but keeps going on. On the contrary, thermal power plants using oil or natural gases are of flexible type in which by just reducing the amount of fuel, the amount of output heat can be reduced. Electricity produced by natural resources, and renewable energies are of stiff type. Thus there is a need of Smart Grid for the human race to survive. Before fossil fuels are exhausted, electricity generation must be transferred to natural energies since there will never come the time of Prometheus III, in which there is no other power supply and the fuel just re-creates itself in larger quantity without any other aids.

In such a case as the Fukushima NPP No 1, the ECCS (Emergency Core Cooling System) is to be ignited to stop the reaction. If circulating water is not supplied, then the temperature of the fuel rods reaches 2800 °C in a few hours and (core) **meltdown** starts.

Even in the situation where meltdown does not occur, if the temperature of fuel pellets comes near 1200°, then the zircalloy–zirconium alloy– contained in the covering tubes of pellets reacts with water vapor to produce hydrogen gas. If hydrogen gas touches oxygen, then *hydrogen explosion* occurs. This really happened in Fukushima. Here's the **peculiarity of hydrogen which is the lightest atom and can penetrate into any small slits**. This peculiarity of hydrogen turns out to be the main ingredient in the hydrogen highway therapy in [Sena and Ohta (2007)]. Thus we are led to the study on hydrogen circulation as well as carbon and nitrogen circulations, which will be done elsewhere.

6.2 Reservoirs of carbons

There are three reservoirs of carbon in the earth system: atmosphere, ocean and land. The amount of reserved carbon varies according to the reservoirs.

(i) Atmosphere

Most of carbons in atmosphere is Carbon dioxide CO_2, then methane which is in much smaller quantity. Methane, however, cannot be ignored since its greenhouse effect is significant.

CO_2: 750 Gt.

CH_4 (methane): 3.7 Gt.

Here $Gt = 10^{15}g = 10^9$ tons.

The ocean is the largest reservoir of carbons.

(ii) Ocean

DIC (Dissolved Inorganic Carbon): $1020 + 38100$ Gt, where 1020 is in the surface and 38100 in the depth of the ocean.

DOC (Dissolved Organic Carbon): < 700 Gt.

Marine biosphere: 3 Gt.

Surface sediments (organic matter, $CaCO_3$): 150 Gt. Here surface refers to the bottom of the ocean.

(iii) Land

Land biosphere: 610 Gt.

Soil: 1500 Gt.

6.3 Change of amount of carbons

Connected with the glacial-interglacial periods, the amount of CO_2 varies. In the transitional time of 6000 years from the **last glacial age** (20,000 years ago) (180 ppm) to the **postglacial age** (10,000 years ago) (270 ppm), the amount of carbon dioxide increased by 90 ppm. In the succeeding 10,000 years, it increased by 10 ppm.

(i) The rate of increase of CO_2 in the glacial-interglacial periods: 27×10^{15} g/1000 years.

(ii) The rate of increase of CO_2 by activities of human being: 29×10^{18} g/1000 years.

(iii) The rate of increase of CO_2 by volcanic gases: 3×10^{13} g/1000 years.

name	CO_2 increase
(i) glacial-interglacial periods	27×10^{15} g/1000 years
(ii) human activities	29×10^{18} g/1000 years
(iii) volcanic gases	3×10^{13} g/1000 years

Table 6.1: Increase rate of CO_2

Comparing (i) and (ii), we see that the **rate of increase of CO_2 by human activities is 1000 times as much as that by nature**

$$29 \times 10^{18} \div 27 \times 10^{15} \approx 1000/1000$$

years.

The issue of increase of carbon dioxide lies not only in the amount but also in its speed of increase. Indeed, the earth used to be rather warm e.g. in the **Cretaceous period** with much higher condensation of carbon dioxide.

The Cretaceous period: $146,000,000 \sim 65,000,000$ years ago belongs to the **Mesozoic era** ($245,000,000 \sim 65,000,000$ years ago). It is said that in the Cretaceous period, the climate was far warmer than now and the density of CO_2 was 10 times as much as the present one.

name	years	climate
Mesozoic period	(1)	very warm
Cretaceous period	(2)	10 times as much CO_2
end of the last ice age	$-12,000$	largest area of glacier
best climate period	$-6,000$	Sahara desert with green
medieval warm period	$950 \sim 1320$	marine transregression
little ice age	$1350 \sim 1820$	marine regression

Table 6.2: Various periods and climate therein

Here (1) $-245,000,000 \sim -65,000,000$ and (2) $-146,000,000 \sim -65,000,000$, where $-$ means "years ago".

The rate of exchange of carbons between two reservoirs is called the **carbon flux**. The exchange with large flux is the one between the atmosphere and the ocean which works for a period of several tens to several hundred thousand years. Indeed, the difference between (i) and (ii) above is caused by the release of carbons from the ocean to the atmosphere.

6.4 Time scale of over one million years

On the other hand, at a timescale of over one million years, we think of *atmosphere plus ocean as one unit.* We interpret the change of amount of carbon as caused by the exchange of between this unit and the other unit "solid earth". At this timescale there are two types of carbon exchanges.

(i) **Chemical weathering of rocks → sedimentation as calcine** described as

$$CaSiO_3 + CO_2 \rightarrow CaCO_3 \downarrow + SiO_3 \downarrow . \quad (6.1)$$

Rocks contain cations such as Calcium, Magnesium, etc. which combine with silica SiO_2 to form minerals such as $CaSiO_3$. When carbon dioxide dissolves into water, carbonic acid is formed. It reacts with minerals containing calcium and melts and decomposes them into calcium cation Ca^+ and carbonic acid anion HCO_3^-, which then dissolves into (sea) water. Then those sea creatures involving shells, coral, planktons (which form calcareous shells) create *calcareous shells*, fixing CO_2, and *a fortiori* carbons as **calcium carbonate** $CaCO_3$ or **limestones.**

(ii) **Synthesizing of organic material by planktons or epigeous plants using carbon dioxide and water → sedimentation of their corpse and simultaneously releasing oxygen** described as

$$CO_2 + H_2O \rightarrow CH_2O \downarrow + O_2 \uparrow, \quad (6.2)$$

where we understand by CH_2O any organic material although it signifies carbohydrate. This is the process of producing organic material by epigeous plants using **photosynthesis** and their corpse are buried in the soil, and in the long run get ripened to become fossil fuel.

6.4.1 *Time scale of several* 100,000 *to dozens of years*

In the time scale of several hundred thousand to several tens years, the two units of atmosphere and ocean that are viewed as one in §6.4 must be thought of as separate units and we are to consider the transfer of carbon dioxide between them.

The layer of sea of depth between 100 to 200 m is called **surface sea water** which is often stirred up by wind and the layer deeper than 200 m is called **deep sea water**.

6.4.2 *Three pumps*

There are three pumps working to push carbon dioxide into deep sea water—bio-pump, alkali-pump and solution-pump.

Equation (6.2) in the case of plant planktons shows the process of their production of carbohydrate by photosynthesis and releasing of oxygen. During the process, the carbohydrate including their corpse fall into deep sea water and 99% of them are oxidized to dissolve, releasing carbon dioxide there. This process is called the **bio-pump**, which occurs in surface water. In total, it can be said that by the work of bio-pump, carbon dioxide is pushed *temporarily* into deep sea water.

Although it seems that the carbohydrates produced by plant planktons are so small that they sink very slowly and get dissolved during the sinking, there is a lot of carbohydrate involved in the sediment at the bottom of the sea. This is due to *weight effect* caused by those creatures which feed on plant planktons. They collect the corpse of planktons to make bigger castings so that they sink 100 times faster than in the case where there is no such a creature.

There is a speculation that the appearance of multi-cellar creatures which have calcareous shells and remain as fossil accelerated the fixing of carbon dioxide and thereby caused the mode jumping.

The power of bio-pump is determined by the amount of **nutritive salt** and in particular by that of phosphate supplied to surface water. By comparison of the constituents of planktons collected from various places, it has turned out that the mol ratio is constant, called the **Redfield ratio**:

$$\mathrm{P} : \mathrm{N} : \mathrm{C} = 1 : 16 : 106. \tag{6.3}$$

For nitrogen circulation, cf. [Sena and Ohta (2007)] and for phosphate cf. [Galhotra *et al.* (2015)].

Next we consider the **carbonate pump** which has a converse function to the alkali-pump. It is described by

$$\mathrm{Ca}^{2+} + 2\mathrm{HCO}_3^- \to \mathrm{CaCO}_3 + \mathrm{H}_2\mathrm{CO}_3 \tag{6.4}$$

and takes place in surface sea water. It produces calcium carbonate and releases carbon dioxide into the air.

The inverse process to the carbonate pump (6.4) is called the **alkali pump (in the narrow sense)** and is described by

$$\mathrm{CaCO}_3 + \mathrm{H}_2\mathrm{CO}_3 \to \mathrm{Ca}^{2+} + 2\mathrm{HCO}_3^-. \tag{6.5}$$

This process takes place in deep sea water as follows. When carbon dioxide is pushed into deep water by the bio-pump, it reacts with water to produce **carbonic acid** H_2CO_3 and the water becomes acidic. The alkali pump is the process of dissolving calcium carbonate by carbonic acid to get neutralized. The deep water being neutralized does not return carbon dioxide to the air when it **upwells** to the surface.

These three processes–bio-pump, carbonate pump, and the alkali pump in the narrow sense–altogether push carbon dioxide into deep sea water and is called the **alkali pump** in the wide sense.

The third pump–the solution pump–is the solubility of carbon dioxide into water according to the temperature. The colder the water, the more soluble the carbon dioxide. In the glacial period, it is estimated that the temperature in deep sea water went down by 3 °C, so that 40 Gt carbon dioxide was absorbed into the ocean.

6.4.3 *An analogue of the carbonate pump*

As an illustration of the kingdom of ancient people we consider the **mortar** which has been used as one of the fundamental building material from prehistoric times. It is also called (lime) plaster or stucco.

To produce it one calcines calcium carbonate or limestones $CaCO_3$ at 900 °C to make it into **calcium oxide** (also called calcined lime, quick lime) CaO. Then add water to make it into **calcium hydroxide** $CaOH_2$. Finally adding glue and fiber one makes mortar. When pasted on the wall, the mortar reacts with carbon dioxide in the air to start hardening to go back ultimately to the state of calcium carbonate.

In chemical expressions the above process may be stated as calcining process

$$CaCO_3 \to CaO + CO_2 \uparrow . \tag{6.6}$$

calcium hydroxide producing process

$$CaO + H_2O \to CaHO_2. \tag{6.7}$$

hardening process

$$CaHO_2 + CO_2 \to CaCO_3 + H_2O. \tag{6.8}$$

6.5 Global warming

As we have seen in §6.3, the extraordinary speed of increase of the amount of carbon dioxide is the grave issue. Since the ocean is the biggest reservoir of carbons and it is already warmed up, the temperature will keep increasing even if we can stop releasing even a bit of carbon dioxide on the ground that warmed water in a large vessel remains warm longer than that in a smaller vessel. There is a prediction that even if we stop all release of the carbon dioxide, the increase will continue in the coming 50 years.

Another concern is that from the geo-data, it is seen that in the dynamical system of climate there are a few stable modes in climate and that by some outer cause, the mode jump occurs, then the new mode will be kept even as the outer conditions go back to the previous state, i.e. **mode jumping**.

The third problem is that under drastic climate change, we cannot adopt the old method of human beings moving to a new land. Since the major living areas where one can live a cultural life is already exhausted and there is no more new land to move. This makes global warming an international affair that the human race is to confront.

Although global warming should be a *global* problem for all the human race, it is usually the case that the developing countries are releasing more carbon dioxide to develop their industries, following the same track that the developed countries once went along, and there is no global understanding. There are many reasons for this type of let-it-be-as-others'-problem. One is that it cannot be seen very clearly and people tend to leave such things to a later date as needless fear. Secondly there is a misleading information scattered around which ignore the time scale. The mass communication reports are often more misleading because they agitate and exaggerate tiny points and mix up real data with false illusion.

The period 950–1250 is called the **medieval warm period** while 1400–1700 is called the **little ice age**. We note, however, that between these two periods, the difference of temperature on the whole earth is only 0.4 °C! Therefore, the climate change is rather the change of its spatial pattern than the change of average temperature. I.e. both in medieval warm period and little ice age, there used to be both warmer and colder areas and in general in the latter period, it was colder than in the former.

In conclusion, that at present the carbon flux is 1000 times quicker than that by nature means that by 1000 times quicker the amount of carbon dioxide in the atmosphere will increase and will give rise to a similar hot climate as in the Cretaceous period.

6.6 Dansgaard-Oeschger cycle

In this subsection further study is made on the Dansgaard-Oeschger cycle (D-O cycle) and Heinrich event etc. It is published in Nature in 1993 and refers to a rapid climate fluctuations with extremely large amplitude that occurred 25 times during the last glacial period (75000–20000 years ago).

In the Northern Hemisphere, the course of the D-O cycle takes the form of rapid warming periods whose length are of the order of decades, and each warming period is followed by gradual cooling period which lasts a few hundred years. A 5 °C change over 30–40 years is common in average annual temperatures on the Greenland ice sheet. On the other hand, during the cold period there is an expansion of the polar front, with ice floating further down to the south across the North Atlantic Ocean.

Some scientists say that the D-O cycles occur quasi-periodically with a period length being a multiple of 1,470 years, but this is still speculative. This length of time reminds us of Nishioka's speculation of climate cycles of 700 years [Nishioka (1948)]. Comparing the graphs, we may conjecture the quasi-periodicity of around 1,470 years.

6.7 Determination of era and temperature

(i) The determination of the era by radioactive isotopes. Cf. [Wada *et al.* (1995)]. The method of using radioactive isotopes to determine the era is well-known along with that using radioactive material including Polonium. One of the most famous stories is the determination of a picture to be genuine or not.

The method based on radioactive isotope ^{14}C has been widely used and produced important data on climatology. However, it is speculated after the nuclear experiments conducted in the 1950's changed the ratio of the amount of ^{14}C and the method has more deviations than before. Carbon has two stable isotopes ^{12}C, ^{13}C, and the radioactive isotope is ^{14}C. The ratio of their amount in nature is $99 : 1$. ^{14}C is generated by heat neutrons released from cosmic rays and falls down to the earth, getting absorbed in the bodies of living organisms. For an application, cf. Yamamoto [Yamamoto (1976)].

(ii) The determination of temperature in the past. Cf. [Tada (2013)], [Wada *et al.* (1995)]. This is not so well-known as the carbon 14 method. Most of the oxygen is ^{16}O of mass 16. There are other isotopes ^{17}O and ^{18}O which occupy only $\frac{2.4}{100}\%$ of all isotopes of oxygen. When the temperature is lower, the ratio of the lighter oxygen isotope ^{16}O increases since it vaporizes into the air and gets frozen in snow form. The ratio of the lighter oxygen isotope in the ice-core and snow crystal is the same. Since the fluctuation of $\delta^{18}O$ is connected with the Milankovitch cycle, one can determine the water temperature.

We consider the phase change

$$^{I}X \to {}^{I}Y, {}^{II}X \to {}^{II}Y, \tag{6.9}$$

where the prefix indicates different atomic numbers. We let the reaction constant to be k_1 and k_2, respectively. Also consider the reverse reaction to (6.9) with reaction constants k_1^- and k_2^-, respectively:

$$^{I}Y \to {}^{I}X, {}^{II}Y \to {}^{II}X, \tag{6.10}$$

and consider a possible reaction

$$^{I}X + {}^{II}Y \to {}^{II}X + {}^{I}Y, \tag{6.11}$$

whose reaction speed to the right is v_1 and v_2 to the left and think of it being in equilibrium, i.e. $v_1 = v_2$. Then

$$v_1 = k_1[{}^{I}X]k_2^-[{}^{II}Y] = k_1k_2^-[{}^{I}X][{}^{II}Y], v_2 = k_2k_1^-[{}^{II}X][{}^{I}Y] \tag{6.12}$$

with the square bracket indicating the molarity. Recall that the equilibrium constant K is defined by

$$K = \frac{k_2k_1^-}{k_1k_2^-}. \tag{6.13}$$

Equating v_1 and v_2 in (6.12), we have

$$K = \frac{k_2k_1^-}{k_1k_2^-} = \frac{[{}^{I}X][{}^{II}Y]}{[{}^{II}X][{}^{I}Y]}. \tag{6.14}$$

Recall that the **isotope ratio** r_Y of the substance in the state Y is defined by

$$r_Y = \frac{[{}^{II}Y]}{[{}^{I}Y]}. \tag{6.15}$$

Using (6.15) in (6.14), we may express the fractionation constant $\alpha_{X,Y}$ as

$$\alpha_{X,Y} = K = \frac{r_Y}{r_X}. \tag{6.16}$$

We specify the above reactions with stable isotopes of oxygen. (6.9) reads

$$H_2{}^{16}O_{(aq)} \to H_2{}^{16}O_{(g)}, H_2{}^{18}O_{(aq)} \to H_2{}^{18}O_{(g)} \tag{6.17}$$

with reaction constants k_1, k_1^-, k_2, k_2^- and (6.11) reads

$$H_2{}^{16}O_{(aq)} + H_2{}^{18}O_{(g)} \leftrightarrow H_2{}^{18}O_{(aq)} + H_2{}^{16}O_{(q)} \tag{6.18}$$

with constants v_1, v_2. Note that

$$\begin{aligned} v_1 &= k_1k_2^-[{}^{I}X][{}^{II}Y] = k_1k_2^-[{}^{16}O_{(aq)}][{}^{18}O_{(g)}], \\ v_2 &= k_2k_1^-[{}^{II}X][{}^{I}Y] = k_1k_2^-[{}^{18}O_{(aq)}][{}^{16}O_{(g)}]. \end{aligned} \tag{6.19}$$

Equating $v_1 = v_2$, we have

$$\alpha_{aq,g} = \frac{k_2 k_1^-}{k_1 k_2^-} = k_1 k_2^- [^I X][^{II} Y] = \frac{[^{16}O_{(aq)}][^{18}O_{(g)}]}{[^{18}O_{(aq)}][^{16}O_{(g)}]}, \tag{6.20}$$

where $\alpha_{aq,g}$ is the **isotopic fractionation constant.**

(6.15) reads

$$r_{O_{(g)}} = \frac{[^{18}O_{(g)}]}{[^{16}O_{(g)}]}. \tag{6.21}$$

Hence it follows that

$$\alpha_{aq,g} = \frac{r_{O_{(g)}}}{r_{O_{(aq)}}}. \tag{6.22}$$

In the analysis of the Greenland's ice-core, the amount of $\delta^{18}O$ was analyzed and determined the temperature at each period.

Stable isotope compositions of low-mass (light) elements such as oxygen, hydrogen, carbon, nitrogen, and sulfur are normally reported as δ values. δ values are reported in units of parts per thousand (denoted as or per mil, or per mille — or even recently, per mill) relative to a standard of known composition. δ values are calculated by

$$\delta \quad (\text{in } 0/00) = \left(\frac{r_x}{r_S} - 1 \right) \times 1000, \tag{6.23}$$

where r denotes the ratio of the heavy to light isotope (as in (6.15)), and r_x and R_S are the ratios in the sample and standard, respectively.

6.8 Weather cycle conjecture: Multiple of ∼700

Figure 6.1: The temperature record found in Greenland ice-core in 1993.
Figure 6.2: The temperature record mainly in Taiwan found in dendroclimatology due to H. Nishioka, 1948.
Figure 6.3: Comparison of Figures 6.1 and 6.2.

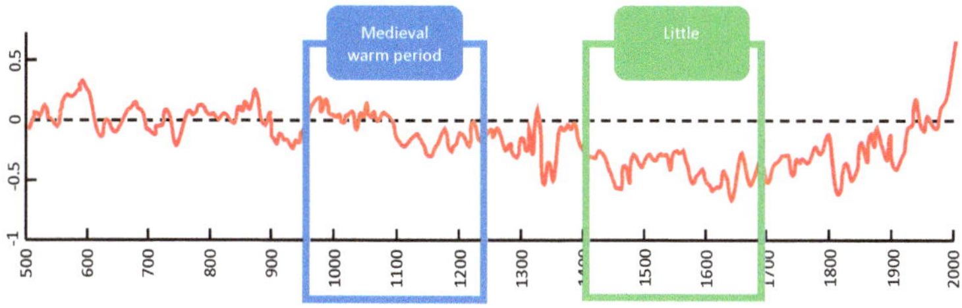

Fig. 6.1: Greenland ice-core graph

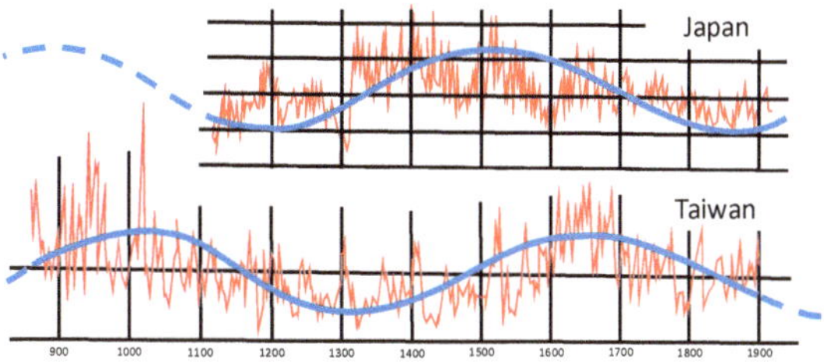

Fig. 6.2: Nishioka graph

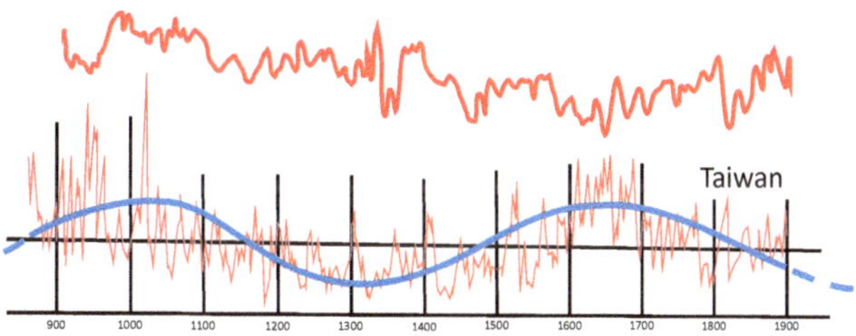

Fig. 6.3: Greenland-Nishioka graph

Looking at Fig. 6.3, we are tempted to speculate a plausible weather-cycle conjecture as a combination of the latest conjectures based on the Dansgaard-Oeschger (DO) cycles and the conjecture of H. Nishioka:

Weather cycle is a multiple of ∼700 years.

But the 1000 times speed of increase of CO_2 stated in §6.3 may affect this.

Chapter 7

EVs in Green Grid

A very plausible picture of the near future–some decades from now–would be the **Green Grid Era** and the emerging importance of EVs in the spectrum, where by Green Grid we mean the advanced Smart Grid in which the essential part of the whole electric power is supplied by distributed generation, i.e. it is generated by renewable energy including solar energy, biogas, geothermal energy, etc. The crucial ingredient in this grid system are enhanced batteries and the participation of EVs in the grid as a pool of electricity. An EV will be like a **smart phone** which works as a vehicle. Once a driver gets on his/her EV, it gets connected to GG by wireless communications system and GG will control it once the destination is input.

The biggest defect of electricity is that it **cannot be stored** in a large quantity and therefore must be kept producing, which is one of the two main excuses of big power companies for their bulldozing plans of building new power stations including nukes. The only major storage unit in most power stations is the **pumped storage systems**, i.e. in the form of potential energy. Electricity can be stored in batteries (or super-capacitors) only in small quantity.

The main ingredient of the cost of electricity is the cost of the high-voltage power lines, which is composed of a grid (system). In general, the operation cost of a grid depends highly on the **PAR (peak-to-average ratio)** in aggregate load demand. For example, there is usually at least one major peak in a daily (residential as well as industrial) load demand profile and some 10 peak hours during the whole year.

Here arises the second of two main excuses of big power companies for building new power stations: To assure reliable service including the supply at peak hours, the companies must produce power superseding the peak value. This makes the value of PAR higher and can significantly increase the generation cost since the grid will be **highly underutilized most of the time**.

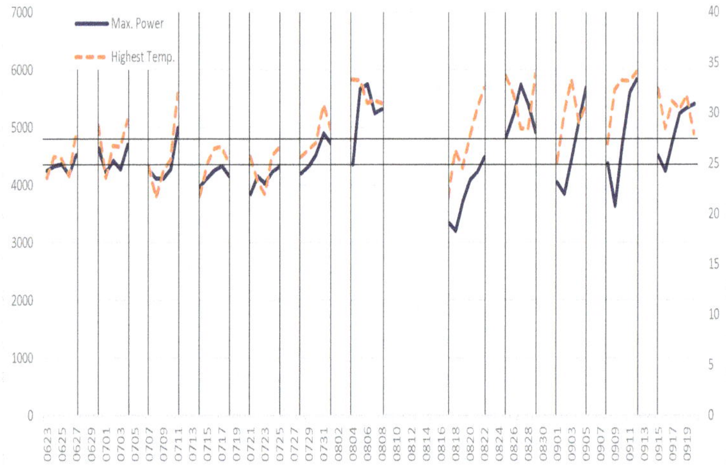

Fig. 7.1: Daily max. temp. and power in Tokyo, 2003

7.1 Future cars will all be EVs

It is certain that all fossil fuels will be used up in the near future and any industrial activities will be estimated from the point of view of WTW efficiency **From Well To Wheel efficiency**, i.e. the efficiency of the used material and environmental protection, as opposed to heat efficiency in §2.6. Only those activities will be allowed to make which have high WTW efficiency and are environmental-friendly. As long as fossil fuels are available, inner combustion engine cars can be driven with high cost of damaging the environment. An **electric vehicle** (EV) is a type of a car which moves by transmitting the power from a battery in a car to wheels. The WTW efficiency for inner combustion cars resp. for EVs is the efficiency of oil dug out from the oil well, processed and finally used as gasoline or diesel oil. resp. that of oil used to generate electricity. There is a principle in the universe that the simplest is the best. In this respect, EVs are the strongest candidate for the only possible solution for transportation save for railroads. For the principle of their mechanism is so simple that already in 1873, 12 years before the first inner combustion engine cars, EVs were

	strong pt	defect
electricity	clean if the generation is so	cannot be stored
fossil fuels	can be regulated	increase of CO_2
nukes	looks clean	radiative pollution

Table 7.1: Defects of electricity

produced for practical use. Indeed, it consists of a battery and a motor in place of a very complicated mechanism of inner combustion engines whose heat efficiency is only 12% since they can use only the explosion energy. However, cheap oil prevailed the market and the last 100 years were the time of inner combustion engine cars. It is the time to think of the future when there is no more fossil fuel available. Then we need to appeal to other natural resources, esp. solar energy which is the largest of all and is the source for other natural resources including wind, water etc. EVs can move as long as they have electricity and electricity in turn has a great variety of ways to generate. As will be mentioned, **wireless in-wheel EVs** [Sato *et al.* (2014)] have a possibility of charging during travel on highways from the coils set up on the roadside. This is expected to solve the weak point of EVs that they are expensive because of the batteries–lithium ion batteries in particular.

The future cars must be EVs connected to the **Green Grid** (GG) and are being monitored by the control center of GG. Here we mean GG the advanced Smart Grid which does not depend on atomic energy nor fossil fuels and is to be the cleanest of all.

As mentioned in [Li *et al.* (2015)], electricity looks clean.

In this chapter, we discuss the emerging importance of EVs in the coming GG Era and also mention the global warming as an aspect of climate change. An EV is a vehicle so that vehicle dynamics applies but with flexibility of in-wheel motors and we may think of a fail-safe guard, an advanced collision avoiding system.

What is proposed is the oblique or even lateral drive to avoid danger ahead. Reaction by humans takes some time and this loss of time may result in probable catastrophe. There is an elapse of time of a total 0.05 s for the information transmission through nerves (0.02) together with the mechanical reaction time (0.03) of the vehicle before any action can be taken. Moreover we must recall the startling fact that we live 0.5 s past [Pollack (1999)] (or more recent 80-millisecond rule). Therefore, in the future situation where all EVs are connected to the GG, each ancillary station (real or virtual) is monitoring the positions of all of the vehicles on the map and can send orders to the driving system directly to take a necessary action as soon as the computer can predict the danger. This includes not only sudden brakes but also shifting to the safer side of the road by lateral driving.

7.1.1 *Yonden EV "PIVOT"*

The Yonden EV "PIVOT" has two driving modes. In addition to normal drive mode there is a universal drive mode in the negative and positive phases. The negative (res. positive) phase is the one with the front and rear wheels in opposite (res. the same) direction. The negative phase enables a pivot rotation including a point rotation and the positive phase enables an oblique drive including a lateral drive.

The control system can separately and immediately control the traction force and the steer angle for each wheel by the instruction of the Grid Operator.

phase	positive	negative
wheels	the same	opposite
drive	oblique	pivot
special	lateral	sharp turn

Table 7.2: Yonden EV universal drive

7.2 Fail-safe guard by GG

It is a world tendency for newly sold vehicles to have a collision avoidance system (CAS) (also known as a precrash system) — the system designed to reduce the severity of a collision — including autonomous emergency braking (AEB) and autonomous emergency steering (AES). It uses radar (all-weather) and sometimes laser (LIDAR) and camera (employing image recognition) to detect an imminent crash. GPS sensors can detect fixed dangers such as approaching the stop signs through a location database. Once the detection is done, these systems either provide a warning to the driver when there is an imminent collision or take action autonomously without any driver input (by AES or AES or both). Collision avoidance by AEB is appropriate at lower speeds (e.g. below 50 km/h), while collision avoidance by AES is appropriate at higher speeds. Cars with collision avoidance may also be equipped with adaptive cruise control, and use the same forward-looking sensors.

A wireless In-Wheel Motor EV (IWM) [Sato *et al.* (2014)] is a new kind of EV in which power is transferred to motors by wireless transmission rather than via cables. This will solve the problem of wire disconnection by liquidating the complicated wiring. It also has a remarkable feature that it can be charged while driving.

Indeed, IWM EVs have been known for quite some time [EE (1999)], [Nishiyama *et al.* (2016)] etc. Here we take up the **Yonden EV "PIVOT"** [Nasu and Higasa (1995)] which have realized two driving modes 20 years ago. We are more interested in the advanced — wireless IWM version — of Yonden pivot which is to be a model for all possible wireless IWMEVs as the most essential component of the GG. In the near future when fossil fuel is being exhausted, the ideal vehicles will be Yonden pivot type. The present situation looks still "A tail wagging its dog." EVs have been killed once [EC (2006)] (or the second time counting the moment when they are expelled from the market by gasoline-propelled cars) and although they are being revived [Paine (2011)], the time may not be the most proper. The transient period may last long, up to 40 years, say until all cars will be replaced by EVs because the EVs can compete against conventional cars only when the infrastructure is more complete. In the coming society when the infrastructure is fully developed, all cars will be wireless IWM EVs that are connected wireless-wise to the GG through regional aggregators during moving and while driving on highways they can be charged by the chargers which are charged by solar energy. Since all cars are monitored by the GG in general and by the regional aggregator in particular, the Grid can check the traffic in real time and whenever it detects possible collision, it can send orders directly to the driving system to make a necessary action. In this way, GG can realize a real time collision avoidance system, which does avoid collision rather than reducing the damage by collision.

The GG control system is to be able to monitor the surroundings of each individual EV and immediately control the traction force and the steer angle for each wheel by the instruction sent wirelessly from the Aggregate controller.

We illustrate the collision avoidance when there is an obstacle on the lane ahead by oblique or lateral drive by Yonden type.

Figure 7.2: An EV is under surveillance of Aggregator 1.
Figure 7.3: The EV moves to the region under surveillance of Aggregator 2.
Figure 7.4: The EV moves to the region under surveillance of Aggregator 3.

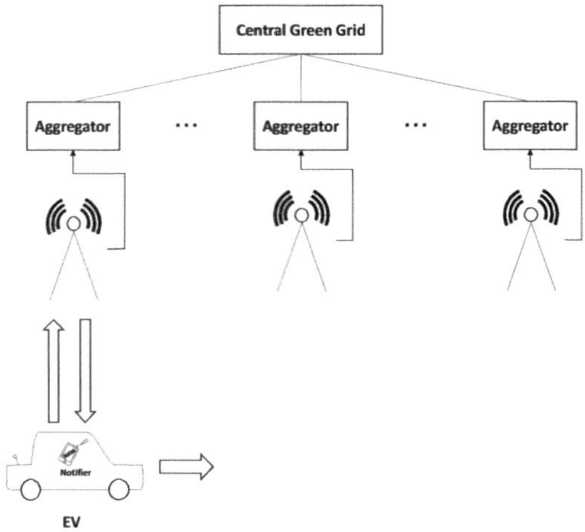

Fig. 7.2: Aggregator 1

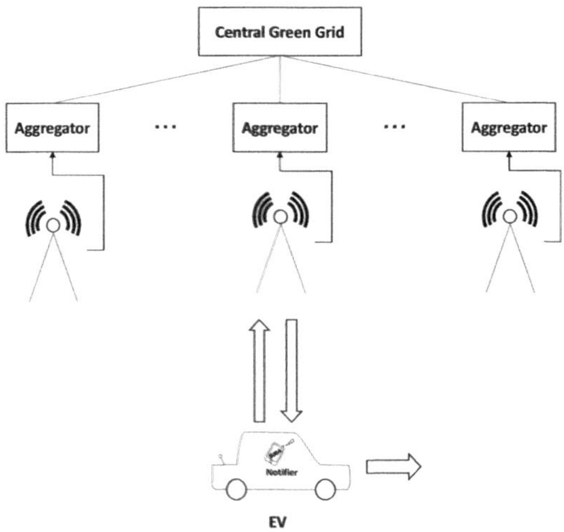

Fig. 7.3: Aggregator 2

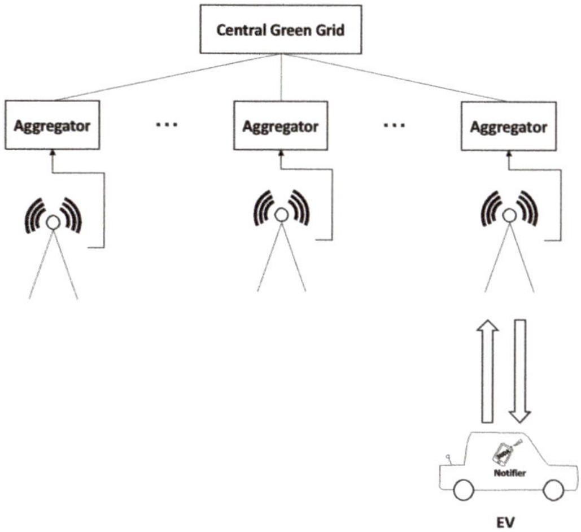

Fig. 7.4: Aggregator 3

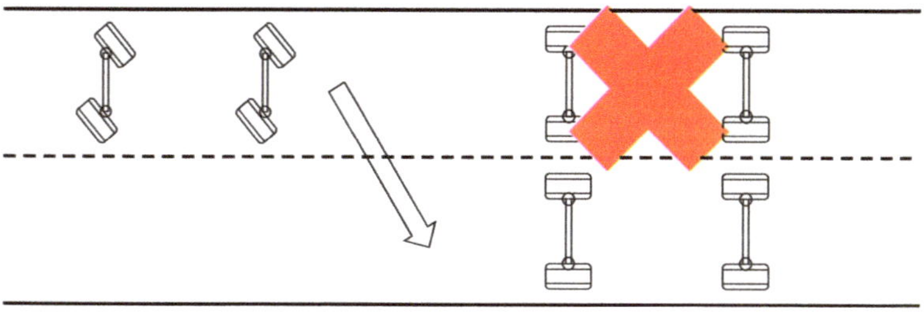

Fig. 7.5: Oblique drive

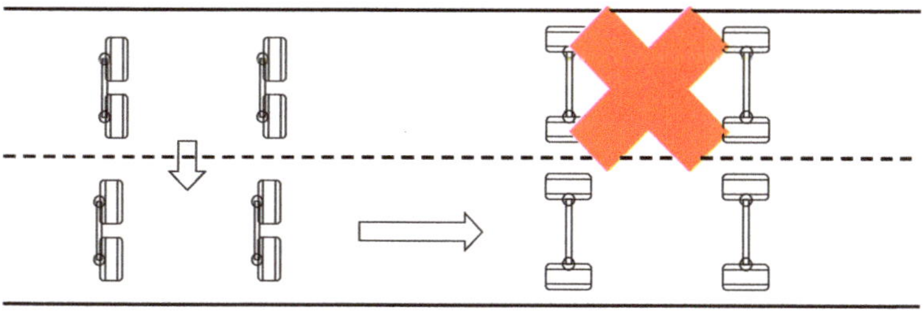

Fig. 7.6: Lateral drive

7.2.1 *Cyberattack impact on GG*

Through wireless communications system, the grid functions may be improved and EV2GG will mitigate many energy and environmental problems. However, *wherever there is light, there is shadow.* The present grid has almost no fear of cyberattack. For a controlled grid, there will be increased vulnerabilities since there will be many components connected including control of many EVs connected to GG, the complexity of integration (of information technology in traditional power grid), diversity of system vendors, urge for timely solutions etc.; all which lead to an **increased risk of cyberattacks**.

We may apply the theory of **graph-based dynamical systems** [Kundurm *et al.* (2014)], [Feng *et al.* (2010)]. A dynamical system paradigm gives a flexible framework to model the cause-effect relationships between the cyber data and electric grid states signals.

Another possible menace will come from using conventional crypto-systems. There is a possibility of realization of quantum computers which will break many crypto-systems, cf. §1.5.3.

Chapter 8

Signal restoration: From Paley-Wiener to Abel continuity theorem

Abstract

In this chapter we state some basic results on signal restoration based on Shannon's sampling theorem which in turn depends on the Paley-Wiener theorem. A systematic exposition based on Fourier analysis is given in [Chakraborty *et al.* (2016), §28, §29]. Although here we reproduce the basic part of Fourier analysis and proof of the sampling theorem therewith, our new contribution is the use of a general form of the Laplace transform due to Widder [Widder (1946)], cf. [Sumi (1999)]. Thus we may say that we give a Laplace transform foundation of the sampling theorem through the Payle-Wiener theorem, which is not surprising if we recall that the Laplace transform is a special form of the Fourier transform.

Indeed, we go on further to find that the missing link among a few crucial results in analysis, Abel continuity theorem, convergence theorem on (generalized) Dirichlet series, and Paley-Wiener theorem is the Laplace transform with Stieltjes integration. By this discovery, the reason is made clear why the domains of Stoltz path and those of convergence look similar. Also as a natural intrinsic property of the Stieltjes integral, the use of partial summation in existing proofs is elucidated. Here we assume the basics of complex analysis in Appendix C [Chakraborty *et al.* (2016)] and Fourier analysis [Kanemitsu and Tsukada (2007)].

8.1 Signal restoration

In Chapter 4 (control theory) and Chapter 7 (EV2GG) of the present book, (wireless) communication is essential. For this the signal being digital, the signal restoration is the main issue. The celebrated sampling theorem (Shannon 1949) being essential in frequency analysis and in particular, in signal transmission, guarantees the restoration of the digital signal by sampling. We state it here as Theorem 8.1 to motivate our analysis and we will come back to this as Theorem 8.6 below. For terminology cf. §8.2.

Theorem 8.1. *If the function $f(t)$ is band-limited with band-length 2Ω:*

$$\hat{f}(\omega) = 0 \quad |\omega| \geq \Omega > 0, \tag{8.1}$$

then f is uniquely determined by its values (samples) at a sequence of equidistant points, by the Nyquist rate $\Delta t = \frac{1}{2\Omega}$ apart:

$$f(t) = \sum_{n=-\infty}^{\infty} f_n \operatorname{si}(2\pi\Omega(t - n\Delta t)) = \sum_{n=-\infty}^{\infty} f(n\Delta t)\frac{\sin(2\pi\Omega(t - n\Delta t))}{2\pi\Omega(t - n\Delta t)}, \tag{8.2}$$

where

$$f_n = f(n\Delta t) = f\left(n\frac{1}{2\Omega}\right) \tag{8.3}$$

is the Nyquist sample.

As an example of the Nyquist rate, we refer to [Kumagai (2007), pp. 30-33], where the principle of CD is stated. The pitch of the sound can be divided into $2^{16} = 65536$ parts because a CD can record 16 convex-concave points as one information and transforms into 0 and 1 signals. The CD reads these 16 information 44100 times per second. The reason for this depends on the assumption that human ears can hear the sound whose frequencies are up to 20 kHz = 20000 Hz. Since the sound with frequency 20000 Hz, by definition, oscillates 20000 times per second and so more than this number of sampling is needed. And for stereo recording, we need twice as many, whence the sampling frequency 44100. Thus the sampling (Nyquist) rate $\frac{\pi}{T}$ is 1/44100 and the input digital signal is restored by the sampling theorem above.

The assumption that all the frequencies higher than 20000 Hz may be neglected is rather controversial. For recording a simple conversation more may not be needed, but for music as supreme art, this omission can be a serious problem because what is missing is often more meaningful as art. We recall that as soon as the natural real sound is transformed into digital signals, it is not the real sound but an approximation.

In scientific disciplines, there are some empirical *threshold numbers* which are used often with great efficiency, although the reason why such thresholds appear is not theoretically clear.

The following theorem ([Weaver (1983), Theorem 5.2, pp. 119-120]) explains why the sampling rate may be taken as the **10 samples per cycle of the highest frequency**.

Theorem 8.2. (Real world sampling theorem) *If the function $f(t)$ is almost band-limited with band-length 2Ω and* **almost time-limited** *with time limit $2X$:*

$$\int_{-\infty}^{X} |f(t)|\,dt < \varepsilon, \quad \int_{X}^{\infty} |f(t)|\,dt < \varepsilon \tag{8.4}$$

then $f(t)$ can be recovered to any desired accuracy from its sampled values at uniformly-spaced intervals $|\Delta t| < \frac{1}{10\Omega}$ apart:

$$f(t) = \sum_{n=-M}^{M} f(n\Delta t) \frac{\sin(2\pi 5\Omega(t - n\Delta t))}{2\pi 5\Omega(t - n\Delta t)} + R_\varepsilon(t), \tag{8.5}$$

where

$$|R_\varepsilon(t)| \ll \varepsilon, \quad M\Delta t > X, \quad \Delta t < \frac{1}{10\Omega}. \tag{8.6}$$

Remark 8.1. An example of a function which is both almost band-limited and almost time-limited is the Gaussian function or its several-variables version

$$f(t) = e^{-at^2}, \quad a > 0. \tag{8.7}$$

This is a density function of the normal (or Gaussian) distribution as well as an almost unique example of a rapidly decreasing function. It appears in the modular relation theory [Kanemitsu *et al.* (2016)] as the intrinsic function to the celebrated Riemann zeta-function.

Thus many research fields are interrelated in a hidden way through the Gaussian distribution.

Since the exponential function may be thought of as a degenerated form of the Bessel functions, one may think of a generalization of the presented ideas with the Bessel function instead.

8.2 Paley-Wiener theorem from Fourier analysis

In this section we explain the Paley-Wiener theorem, Theorem 8.3, and its application to signal transmission from the point of Fourier analysis. First we prepare the relevant function. The Taylor expansion

$$\frac{\sin z}{z} = \sum_{n=0}^{\infty} (-1)^n \frac{z^{2n}}{(2n+1)!}$$

is true for $z \neq 0$.

Defining the value at $z = 0$ of the function $\frac{\sin z}{z}$ to be 1 (i.e. removing the removable singularity), the new function **sinus cardinalis**, often denoted by $\mathrm{si}(z)$, is an entire function:

$$\mathrm{si}(z) = \left(\mathrm{sinc}\left(\frac{z}{\pi}\right)\right) = \sum_{n=0}^{\infty} (-1)^n \frac{z^{2n}}{(2n+1)!} = \begin{cases} \frac{\sin z}{z} & (z \neq 0) \\ 1 & (z = 0) \end{cases}. \tag{8.8}$$

The modified function $\mathrm{sinc}(t) = \mathrm{si}(\pi t)$ is also often used since the system of sinc-functions forms an ONS (cf. [Chakraborty *et al.* (2016), Theorem 9.1, p. 62]).

8.2.1 *Fourier analysis*

We make a brief review on Fourier analysis based on the Lebesgue integration. Cf. DFT in Definition 1.4 and the Fourier series in (4.73) for periodic functions.

Let $L^p(D)$ denote the **(Banach) space of all pth power integrable functions** (in the Lebesgue sense), where D indicates the real line $\mathbb{R}$ or an interval $I \subset \mathbb{R}$ with p-norm

$$\|f\|_p = \left(\frac{1}{\sqrt{2\pi}}\int_D |f(t)|^p \,\mathrm{d}t\right)^{1/p} \tag{8.9}$$

and $\|f\|_\infty = \lim_{p\to\infty}\|f\|_p = \sup_{t\in\mathbb{R}}|f(t)|$. For any function $f \in L^1$, we define its **Fourier transform**

$$F(\omega) = \hat{f}(\omega) = \frac{1}{\sqrt{2\pi}}\int_{-\infty}^{\infty} f(t)e^{-i\omega t}\,\mathrm{d}t = \lim_{T\to\infty}\frac{1}{\sqrt{2\pi}}\int_{|t|\le T} f(t)e^{-i\omega t}\,\mathrm{d}t \tag{8.10}$$

for each frequency $\omega \in \mathbb{R}$. By this we go into the space of frequencies:

time domain $\to$ frequency domain.

If $f \in L^2$, the limit in (8.10) is to be understood in the L^2-norm sense.

The underlying principle that supports both approaches by the sampling and Bernstein polynomials is the Paley-Wiener theorem. There are some books and papers related to this theorem (cf. e.g. [Kakita (1985), pp. 38-40]).

Let

$$f(z) = \frac{1}{\sqrt{2\pi}}\int_{-A}^{A} F(\omega)e^{i\omega z}\,\mathrm{d}\omega, \tag{8.11}$$

where $A > 0$ and $F \in L^2(-A, A)$. Then one can show that f is an integral function. It satisfies the growth condition

$$|f(z)| \le \frac{1}{\sqrt{2\pi}}\int_{-A}^{A} |F(\omega)|e^{-\omega y}\,\mathrm{d}\omega \le \frac{1}{\sqrt{2\pi}}e^{A|y|}\int_{-A}^{A} |F(\omega)|\,\mathrm{d}\omega, \tag{8.12}$$

where we write $z = x + iy$. Hence denoting the last integral by C, we deduce (8.43).

Definition 8.1. An integral function f that satisfies condition (8.43) is said to be **of exponential type** (or of order 1 a là Hadamard).

Thus we have

Proposition 8.1. *Every f of the form* (8.11) *is an integral function which satisfies* (8.43) *and whose restriction is in L^2 (by the Plancherel theorem).*

The remarkable theorem of Paley-Wiener asserts that the converse also holds.

Theorem 8.3. (Paley-Wiener theorem) *Suppose A and C are positive constants, that f is an integral function of exponential type, i.e. satisfying* (8.43) *for all values of z, and that*

$$\int_{-\infty}^{\infty} |f(z)|^2 < \infty. \tag{8.13}$$

Then there exists an $F \in L^2(-\infty, \infty)$ such that

$$f(z) = \frac{1}{\sqrt{2\pi}} \int_{-A}^{A} F(\omega) e^{i\omega z} \, \mathrm{d}\omega \tag{8.14}$$

for **all** *values of z.*

The following theorem is fundamental in harmonic analysis.

Theorem 8.4. (Carlson's theorem) *Let $f \in L^p(\mathbb{R})$ for some $p \in (1, \infty)$ have the Fourier transform $F(\omega) = \hat{f}(\omega)$. Then the* **Fourier inversion formula** *holds true:*

$$\frac{1}{\sqrt{2\pi}} \int_{-\infty}^{\infty} \hat{f}(\omega) e^{it\omega} \, \mathrm{d}\omega = \lim_{T\to\infty} \frac{1}{\sqrt{2\pi}} \int_{|\omega|\le T} F(\omega) e^{it\omega} \, \mathrm{d}\omega = f(t) \tag{8.15}$$

for **almost every** *(a.e.) $t \in \mathbb{R}$.*

I.e. *frequency domain $\to$ time domain.*

The discrete analogue of Theorem 8.4 gives the Fourier expansion

Theorem 8.5. (Discrete Carlson's theorem) *Let f be periodic of period 2π and $f \in L^p(\mathbb{R})$ for some $p \in (1, \infty)$ with Fourier coefficients $\hat{f}(n)$. Then the* **Fourier expansion** *holds true:*

$$\frac{1}{\sqrt{2\pi}} \sum_{n=-\infty}^{\infty} \hat{f}(n) e^{int} = \lim_{N\to\infty} \frac{1}{\sqrt{2\pi}} \sum_{|n|\le N} \hat{f}(n) e^{int} = f(t) \tag{8.16}$$

for **almost every** *(a.e.) $t \in \mathbb{R}$.*

We note that this *a.e.* (or often *a.a* = almost all) notion is different from a daily-life interpretation of a.e. The a.e. equality means equality that holds true except for a point set whose measure is 0, e.g. except for the set of all rationals, i.e. except for all numbers that are used daily. With this understanding, we come to know the limitation of the extrapolation etc.

Theorem 8.4 (8.15) then asserts that

$$f(t) = \frac{1}{\sqrt{2\pi}} \int_{-\pi\Omega}^{\pi\Omega} F(\omega)e^{it\omega}\, \mathrm{d}\omega \tag{8.17}$$

holds true for a.a. values of t.

We may "prove" (8.17) by the following lemma.

Lemma 8.1. Dirichlet *The evaluation of the Dirichlet integral reads*

$$\lim_{t\to\infty} \int_{-\infty}^{\infty} \frac{\sin 2\pi tx}{\pi x} \varphi(x)\, \mathrm{d}x = \varphi(0) \tag{8.18}$$

provided that φ is continuous at the origin.

Proof. Divide the integral into three parts $(-\infty, -\varepsilon)$, $(-\varepsilon, \varepsilon)$, (ε, ∞) with $\varepsilon > 0$. Then since the function $\frac{\varphi(x)}{x}$ is integrablel over two infinite intervals, it follows from the Riemann-Lebesgue lemma [Kanemitsu and Tsukada (2007), Proposition 7.1, p. 137] that the corresponding integrals tend to 0 as $t \to \infty$. On the interval $(-\varepsilon, \varepsilon)$, $\varphi(x)$ can be approximated by $\varphi(0)$ and so

$$\begin{aligned} \int_{-\varepsilon}^{\varepsilon} \frac{\sin 2\pi tx}{\pi x} \varphi(x)\, \mathrm{d}x &= \varphi(0) \int_{-\varepsilon}^{\varepsilon} \frac{\sin 2\pi tx}{\pi} \frac{\mathrm{d}x}{x} + o(1) \\ &= \frac{\varphi(0)}{\pi} \int_{-2\pi\varepsilon t}^{2\pi\varepsilon t} \frac{\sin x}{x}\, \mathrm{d}x + o(1), \end{aligned} \tag{8.19}$$

which tends to $\varphi(0)$ in view of

$$\int_{-\infty}^{\infty} \frac{\sin x}{x}\, \mathrm{d}x = \pi. \tag{8.20}$$

For this value, cf. e.g. [Chakraborty *et al.* (2016), p. 23]. □

Heuristic proof of (8.17). Substituting (8.10) for $F(\omega)$ on the left of (8.17), we obtain

$$\frac{1}{\sqrt{2\pi}} \int_{-\pi\Omega}^{\pi\Omega} F(\omega)e^{it\omega}\, \mathrm{d}\omega = \frac{1}{2\pi} \int_{-\infty}^{\infty} f(\tau)\, \mathrm{d}\tau \int_{-\pi\Omega}^{\pi\Omega} e^{i(t-\tau)\omega}\, \mathrm{d}\omega \tag{8.21}$$

on changing the order of integration, which is not assured to hold, thus the proof is heuristic. The inner integral is $\frac{2\sin(t-\tau)\pi\Omega}{t-\tau}$. Hence by the change of variable $t - \tau = u$, we conclude that

$$\frac{1}{\sqrt{2\pi}}\int_{-\pi\Omega}^{\pi\Omega} F(\omega)e^{it\omega}\,\mathrm{d}\omega = \frac{1}{2\pi}\int_{-\infty}^{\infty} f(t-u)\frac{2\sin\pi u\Omega}{u}\,\mathrm{d}u, \tag{8.22}$$

which is (8.17) in view of Lemma 8.1.

Note that [Splettstösser (1985), (2)] asserts the truth of (8.17) for **each** $t \in \mathbb{R}$ on account of Theorem 8.3. In view of this, Splettstösser contains the validity of (8.17) for all values in the definition of band-limitedness, which is a precursor of Theorem 8.6.

8.2.2 *Restoration of signals*

Definition 8.2. Any $f \in L^p$ whose Fourier transform has a support (the domain where it is $\neq 0$) $[-T, T]$ is called **band-limited** with highest frequency T or of **band width** $2T$, i.e. the higher frequency is chopped off, which is often expressed as $2\pi\Omega$ or similarly. The sampling rate $\frac{2\pi}{2T} = \frac{1}{\Omega}$ is called the **Nyquist rate** and the sequence obtained using this rate is called **Nyquist samples**.

The sampling theorem is of utmost importance in signal transmission and in a sense an orthogonal expansion w.r.t. the system of sinc-functions. There are many references on it, cf. Papoulis [Papoulis (1962), pp. 50-52], [Weaver (1983), pp. 117-120] etc.

The notation is slightly different from one to another. T in [Papoulis (1962), pp. 50-52], W in [Splettstösser (1985)], and Ω in [Weaver (1983), pp. 117-120]:

[Papoulis (1962), pp. 50-52]	[Splettstösser (1985)]	[Weaver (1983), pp. 117-120]
T	W	Ω

$$T = \pi W = 2\pi\Omega.$$

The following version is from Papoulis [Papoulis (1962), pp. 50-52], cf. Theorem 8.1 above.

Theorem 8.6. (Sampling theorem) *If the signal $f(t)$ is band-limited, then f is given by*

$$f(t) = \sum_{n=-\infty}^{\infty} f_n \operatorname{sinc}\left(\frac{T}{\pi}\left(t - \frac{\pi}{T}n\right)\right) = \sum_{n=-\infty}^{\infty} f_n \frac{\sin(Tt - \pi n)}{Tt - \pi n}, \tag{8.23}$$

where

$$f_n = f\left(n\frac{\pi}{T}\right) \tag{8.24}$$

is the Nyquist sample, i.e., f is uniquely determined by its values (samples) at a sequence of equi-distant points, by the Nyquist rate $\frac{\pi}{T}$ apart.

Proof. We recall the Fourier expansion of $\hat{f}$ in the interval $(-T, T)$ in Theorem 8.5 in slight different notation, where f_n is $\frac{T}{\pi}c_n$ with c_n the nth Fourier coefficient

$$\hat{f}(z) = \frac{\pi}{T}\sum_{n=-\infty}^{\infty} f_n e^{-in\frac{\pi}{T}z},$$

so that we may write

$$\hat{f}(z) = r_{2T}(z)\frac{\pi}{T}\sum_{n=-\infty}^{\infty} f_n e^{-in\frac{\pi}{T}z}$$

with the rectangular pulse function r_{2T}. Recalling the pair

$$r_{2T}(z)e^{-in\frac{\pi}{T}z} \leftrightarrow \frac{T}{\pi}\mathrm{si}(Tt - n\pi)$$

in [Chakraborty *et al.* (2009), (A.31), p. 217], we immediately conclude (8.23). □

Remark 8.2. Theorem 8.6 has an interesting history as is described in [Splettstösser (1985)]. We shall briefly state some highlights. It was stated in his 1949 paper [Shapiro (1953)], which had been submitted in 1940, and he referred to it as **a result of common knowledge** and referred to Nyquist and Whittaker [Whittaker (1915)]. It is therefore sometimes called the Shannon-Whittaker theorem [Ruderman *et al.* (1992)]. In Whittaker the series (8.23) is known as *cardinal series.*

Third Proof of Theorem 8.6. We view (8.23) as a discrete convolution

$$(f * \mathrm{si})(t) = \sum_{n=-\infty}^{\infty} f\left(\frac{\pi}{T}n\right)\mathrm{si}\left(T(t - \frac{\pi}{T}n)\right) = \sum_{n=-\infty}^{\infty} f\left(t - \frac{\pi}{T}n\right)\mathrm{si}\left(T\frac{\pi}{T}n\right). \tag{8.25}$$

Then noting that $\mathrm{si}(x) = 0$, $x = k\pi, k \neq 0$, $\mathrm{si}(0) = 1$, we conclude that the last expression in (8.25) is $= f(t)$.

Definition 8.3. Given two functions f, g, the integral (resp. the sum) of $f(t)g(x - t)$ over the whole domain of t is called their continuous (resp.

discrete) **convolution** and denoted by $(f * g)(x)$. E.g. if $D = (a, b)$, we define

$$(f * g)(x) = \int_a^b f(t)g(x-t)\,\mathrm{d}t, \quad (f * g)(x) = \sum_{n=a}^{b} f(n)g(x-n). \tag{8.26}$$

If $D = (-\infty, \infty)$,

$$(f * g)(x) = \int_{-\infty}^{\infty} f(t)g(x-t)\,\mathrm{d}t, \quad (f * g)(x) = \sum_{n=-\infty}^{\infty} f(n)g(x-n) \tag{8.27}$$

if the limits exist.

In (8.25), use is made of the commutativity $f * g = g * f$, which is not *a priori* guaranteed, so that the proof is not rigorous as it stands.

8.3 Laplace transform

Let $\{\lambda_n\} \subset \mathbb{R}$ be an increasing sequence for which we may suppose $\lambda_1 > 0$. For a sequence $\{a_n\} \subset \mathbb{C}$, the series

$$f(s) = \sum_{n=1}^{\infty} a_n e^{-\lambda_n s} \tag{8.28}$$

convergent in some half-plane, is called a generalized Dirichlet series.
1. If $\lambda_n = \log n$ with log denoting the principal value, $f(s) = \sum_{n=1}^{\infty} \frac{a_n}{n^s}$ is (an ordinary) Dirichlet series.
2. If $\lambda_n = n$ and $e^{-s} = w$, $f(z) = f(z) = \sum_{n=1}^{\infty} a_n w^n$ is the power series.

In all literature [Hardy (1915)], [Serre (1973)], [Tatuzawa (1980)], etc. the convergence theorem for generalized Dirichlet series, Theorem 8.7 and the Abel continuity theorem, Corollary 8.1 are regarded as independent and proofs are given separately. In [Li *et al.* (2016)] it is shown that Theorem 8.7 entails Corollary 8.1 via a counterpart, Corollary 8.2 together with conformality of the analytic mapping $e^{-s} = w$, thus revealing the reason why the convergence domains are angular domains of a similar shape. The proof uses a general form of the partial summation [Li *et al.* (2016), Lemma 2] for a generalized sequence $\{\lambda_n\}$, thus unifying all existing proofs.

In this section we employ a general treatment by (Lebesgue-) Stieltjes integrals to attain two objects at a stretch. I.e. we follow [Widder (1946)] to introduce Corollary 8.3 whose discrete version leads to Theorem 8.7. In the proof, integration by parts is used which is a more general version of the partial summation. Then on one hand we cover Abel continuity theorem by the convergence theorem, Corollary 8.3, for Laplace transforms and conformality, revealing the reason why convergence domains being similar.

On the other hand, we shall show that the basic part of the Paley-Wiener theorem (cf. e.g. [Kakita (1985)]) is laid by the Laplace transform method. Then we appeal to two fundamental results, the Plancherel formula and the Fourier inversion formula. By finding this hidden link of Laplace transform, we are able to treat these two far-looking objects of Paley-Wiener theorem and Abel continuity theorem in a unified way, up to some auxiliary fundamental results. The Paley-Wiener theorem has recently been highlighted in view of its essential application to signal restoration. In both well-known approaches by sampling [Papoulis (1962)], [Weaver (1983)] and by Bernstein polynomials [Giardina (1984)] the Paley-Wiener theorem plays a fundamental role.

Theorem 8.7. *If the series* (8.28) *is convergent for* $s = s_0 = \sigma_0 + it_0$, *then* $f(s)$ *is uniformly convergent in the right half-plane* $\sigma > \sigma_0$ *in the wide sense and represents an analytic function there. More precisely, let* D *be an angular domain*

$$\sigma - \sigma_0 \geq 0, \quad \arg(s - s_0) \leq \delta \tag{8.29}$$

with $0 < \delta < \frac{\pi}{2}$. *Then* $f(s)$ *is uniformly convergent on* D *in the wide sense.*

Corollary 8.1. (Abel continuity theorem) *Suppose a power series* $f(z) = \sum_{n=1}^{\infty} a_n z^n$ *converges at the point* z_0 *on its circle of convergence. Draw two chords (inside the circle) that start from* z_0 *and form an angle* δ *with the tangent at* z_0 *of the circle* ($0 < \delta < \frac{\pi}{2}$)*. Let* Δ *be the (closure of) intersection of this angular subdomain and the disc of convergence. Then* $f(z)$ *approaches* $f(z_0)$ *as* $z \to z_0$ *in the angular domain inside* Δ*. This is often said as* z *approaches to* z_0 **along Stoltz path**.

Corollary 8.2. (Counterpart of Abel continuity theorem) $f(s)$ *approaches to* $f(s_0)$ *as* $s \to s_0$ *in the angular domain* (8.29).

Lemma 8.2. (i) *The Stieltjes integral* $\int_a^b f\,\mathrm{d}g$ *exists if* f *is continuous and* g *is of bounded variation and linear in* f *and* g*. The role can be changed in view of Item* (ii)*. It holds that*

$$\int_a^b \mathrm{d}g(x) = g(b) - g(a). \tag{8.30}$$

(ii) *The formula for integration by parts holds true:*

$$\int_a^b f(x)\,\mathrm{d}g(x) = [f(x)g(x)]_a^b - \int_a^b g(x)\,\mathrm{d}f(x), \tag{8.31}$$

provided that f is continuous and g is of bounded variation or g is continuous and f is of bounded variation.
(iii) *If g is a step function with jumps a_n at x_n, the Stieltjes integral reduces to the sum:*

$$\int_a^b f(x)\,\mathrm{d}g(x) = \sum_{a<x_n\le b} f(x_n)a_n. \tag{8.32}$$

(iv) *If $f(x)$ is continuous, $\varphi(x) \in L = L^1$ on $[a,b]$ and*

$$g(x) = \int_c^x \varphi(u)\mathrm{d}u \quad x, c \in [a,b],$$

then

$$\int_a^b f(x)\mathrm{d}g(x) = \int_a^b f(x)\varphi(x)\mathrm{d}x = \int_a^b f(x)g'(x)\mathrm{d}x, \tag{8.33}$$

where the last integral exists as a Lebesgue integral.

Lemma 8.3. *Let*

$$f(s) = \int_0^\infty e^{-sx}\mathrm{d}g(x) \tag{8.34}$$

and

$$h(u) = \int_0^u e^{-s_0x}\mathrm{d}g(x) \quad u \ge 0. \tag{8.35}$$

If

$$\limsup_{0\le u} |h(u)| = M < \infty \tag{8.36}$$

with $s_0 = \sigma_0 + it_0$, then (8.34) *is convergent at s_0 and*

$$\int_0^\infty e^{-sx}\mathrm{d}g(x) = (s-s_0)\int_0^\infty e^{-(s-s_0)x}\mathrm{d}h(x) \tag{8.37}$$

the integral on the right being absolutely convergent.

Proof. By Lemma 8.2, we have successively for $X < Y$

$$\int_X^Y e^{-sx}\mathrm{d}g(x) = \int_X^Y e^{-(s-s_0)x}\mathrm{d}h(x) \tag{8.38}$$

$$= \left[e^{-(s-s_0)x}h(x)\right]_X^Y + (s-s_0)\int_X^Y e^{-(s-s_0)x}h(x)\mathrm{d}x.$$

For $X = 0$, $Y \to \infty$, the above tends to $\to (s-s_0)\int_0^\infty e^{-(s-s_0)x}h(x)\mathrm{d}x$ where the passage to the limit follows from

$$\left|(s-s_0)\int_0^\infty e^{-(s-s_0)x}h(x)\mathrm{d}\right| \le M|s-s_0|\int_0^\infty e^{-(\sigma-\sigma_0)x}\mathrm{d}x \le M\frac{|s-s_0|}{\sigma-\sigma_0} \tag{8.39}$$

for $\sigma > \sigma_0$. The convergence is absolute (and uniform). □

Remark 8.3. From Lemma 8.3 it follows that the domain of convergence is a half-plane. For our purpose we need stronger results that follow. In most of existing literature on (generalized) Dirichlet series, the above proof is used with integration by parts (8.38) as partial summation.

Theorem 8.8. [Widder (1929), Theorem 4.3, p. 54] *If the integral* (8.34) *is convergent at* $s = s_0$ *and* $H > 0, K > 1$ *are constants, then the integral is uniformly convergent in the domain*

$$D : |s - s_0| \leq K(\sigma - \sigma_0)e^{H(\sigma-\sigma_0)}, \quad \sigma \geq \sigma_0. \tag{8.40}$$

In the proof, (8.38) with $Y = \infty$ is essentially used.

Corollary 8.3. *If the integral*

$$f(s) = \int_0^\infty e^{-sx}\mathrm{d}g(x) \tag{8.41}$$

is convergent at $s = s_0 = \sigma_0 + it_0$ *and* $K = \frac{1}{\cos\delta} > 1$ *where* δ *has the same meaning in Theorem 8.7, then the integral is uniformly convergent in the angular domain*

$$D_{s_0} : |s - s_0| \leq K(\sigma - \sigma_0), \quad \sigma \geq \sigma_0. \tag{8.42}$$

Suppose $f(z)$ is an integral function satisfying the conditions

$$|f(z)| \leq Ce^{A|z|}, \qquad 0 < A \in \mathbb{R} \tag{8.43}$$

and

$$f(x) = o(1) \quad \text{as} \quad z = x \to +\infty. \tag{8.44}$$

For $\alpha \in \mathbb{R}$ let $\Gamma_\alpha(r)$ denote a ray starting from the origin

$$\Gamma_\alpha = \Gamma_\alpha(r) : z = re^{i\alpha}, \quad 0 \leq r < \infty. \tag{8.45}$$

Let

$$P_\alpha = \{s|\, \mathrm{Re}(se^\alpha) > A\}, \tag{8.46}$$

so that P_α is the **rotated right half-plane** obtained from the right half-plane $\{s|\, \mathrm{Re}\, s > A\}$ by rotation $-\alpha$ in the positive direction. Let $\Phi_\alpha(w)$ be defined by

$$\Phi_\alpha(s) = \Phi_\alpha(f, s) = \int_{\Gamma_\alpha} e^{-sz} f(z)\, \mathrm{d}z, \quad s \in P_\alpha. \tag{8.47}$$

In view of (8.45) and (8.46), the Weierstrass M-test applies and therefore the integral (8.47) converges absolutely and represents an analytic function in P_α for every $\alpha \in \mathbb{R}$. Φ_0 is the **Laplace transform**, which is introduced by (1.124).

$$\Phi_0(f, s) = L[f](s) = \int_0^\infty e^{-sx} f(x)\, dx, \quad \sigma : \mathrm{Re}\, s > a. \tag{8.48}$$

The integral (8.48) is convergent at $s = 0$ in view of (8.44). Hence Corollary 8.3 applies and is analytic in the angular domain (8.42).

Lemma 8.4. *Suppose* $0 < \beta - \alpha < \pi$. *Then in the intersection* $P_\alpha \cap P_\beta$, *we have*

$$\Phi_\alpha(f, s) = \Phi_\beta(f, s). \tag{8.49}$$

Proof. We modify the proof in [Rudin (1986), p. 376]. It suffices to prove this in the case $\alpha > 0$, say, and 0 by rotating the configuration by $-\alpha$ and writing α for $\beta - \alpha$. If we prove this case, then we also prove that $\Phi_\alpha(f, s)$ is an integral expression of the Laplace transform. Let

$$s_{\alpha/2} = |s| e^{-\frac{\alpha}{2} i} \tag{8.50}$$

be a ray. Then

$$\mathrm{Re}(se^\alpha) = \mathrm{Re}(s) = |s| \cos \frac{\alpha}{2}.$$

Hence if

$$|s| > \frac{A}{\cos \frac{\alpha}{2}}, \tag{8.51}$$

then $s_{\alpha/2} \in P_\alpha \cap P_\beta$. Hence on the circular arc $C_R : z = Re^{i\theta}$, $0 \le \theta \le \alpha$, we have

$$\begin{aligned} |f(z) e^{-s_{\alpha/2} z}| &= O\left(\exp\left(AR\right) \exp(-r \cos(\theta - \alpha/2))\right) \\ &= O\left(\exp\left(A - |s| \cos(\alpha/2)\right) R\right) \to 0, \quad R \to \infty \end{aligned} \tag{8.52}$$

as long as (8.51). The Cauchy integral theorem applied to the sector with boundaries $[0, R]$, C_R and $z = re^{i\alpha}$, $r : R \to 0$ shows that the sum of integrals along the rays and the integral along the arc is 0. By (8.52), $\Phi_\alpha = \Phi_0$ on the ray (8.50) with (8.51). But then by the consistency theorem, they must coincide in $P_\alpha \cap P_0$. □

Theorem 8.9. *Every* $\Phi_\alpha(f, s)$ *in* (8.47) *is an integral representation of the Laplace transform* (8.48) *for all* $\alpha \in \mathbb{R}$.

Indeed, for $|\alpha| \le \frac{\pi}{2}$, theorem reduces to Corollary 8.3 and Lemma 8.4. And Φ_α coincides with Φ_0 in the angular domain D_{s_0} in (8.42) and in $P_\alpha \backslash D_{s_0}$, Φ_α gives an analytic continuation.

For $\frac{\pi}{2} < |\alpha| \le \pi$, every $\Phi_\alpha(f, s)$ is an analytic continuation of $\Phi_(f, s)$. This also follows from the Schwarz reflection principle with respect to the imaginary axis.

8.3.1 *Proof of Paley-Wiener theorem*

In this section we elucidate the proof of the Paley-Wiener theorem, Theorem 8.3 above, given in Rudin [Rudin (1986)].

Proof of Theorem 8.3. With enough preparation given toward the end of the previous section, the proof now goes along lines of the proof in Rudin. Condition (8.44) is assured by (8.13). The key lies in the proof of

$$\lim_{\delta\to 0}\int_{-\infty}^{\infty} f_\delta(x)e^{-itx}\,\mathrm{d}x = 0, \quad t\in\mathbb{R}, \quad |t|>A, \tag{8.53}$$

where

$$f_\delta(t) = f(x)e^{-\delta|x|}. \tag{8.54}$$

Since $f_\delta \to f$ in L^2-norm, the Plancherel theorem implies that the Fourier transforms of f_δ tends to $\hat{f}$ in L^2-norm. (8.53) shows $\hat{f}$ vanishes in $|t|>A$ and the Fourier inversion formula entails (8.14) for a.a. values. But then both sides being analytic, hence for all values.

Proof of (8.53) follows from Lemma 8.4. For first note that

$$\int_{-\infty}^{\infty} f_\delta(x)e^{-itx}\,\mathrm{d}x = \Phi_0(f_\delta) - \Phi_\pi(f_{-\delta}). \tag{8.55}$$

Then rewrite this as

$$\int_{-\infty}^{\infty} f_\delta(x)e^{-itx}\,\mathrm{d}x = \begin{cases}\Phi_{-\frac{\pi}{2}}(f_\delta) - \Phi_{-\frac{\pi}{2}}(f_{-\delta}) & t>A\\ \Phi_{\frac{\pi}{2}}(f_\delta) - \Phi_{\frac{\pi}{2}}(f_{-\delta}) & t<-A\end{cases} \tag{8.56}$$

Then (8.53) immediately follows, completing the proof.

Bibliography

Ando, M. (2000). Arsenic contaminate of ground water in India and Bangladesh and its health effect, *J. Nat. Inst. Public Health*, **49**, (3), 266-274.

Andrews, J. G. and McLone, R. R. (1976). *Mathematical modeling*, Butterworths, London.

Apostol, T. M. (1957). *Mathematical analysis*, Addison-Wesley, Reading.

Aviram, M. and Eias, K. (1993). Dietary olive oil reduces low-density lipoprotein uptake by macrophages and decreases the susceptibility of the lipoprotein to undergo lipid peroxidation, *Ann. Nutr. Metab.* **37**, 75-84.

Banerjee, M. B., Shandil, R. G., and Katyal, S. T. (1986). A nonlinear theory of hydrodynamic lubrication, *J. Math. Anal. Appl.*, **117**, 48-56.

Bellman R. (1980). *Analytical number theory (an introduction)*, The Benjamin/Cummings Publ. Co. Inc. New York.

Bently, W. F. (1997). Cell balancing considerations for Loithium-ion battery system, *IEEE*.

Bonfiglio, C. and Roessler, W. (2009). A cost optimized battery management system with active cell balancing for Lithium ion battery stacks, *IEEE*.

Braunstein, S. L. and Pati, A. K. (2012). *Quantum information with continuous variables*, Springer Verl., Heidelberg etc.

Brunson, L. R. and Sabatini, D. A. (2009). An evaluation of fish bone char as an appropriate arsenic and fluoride removal technology for emerging regions, *Environ. Engrg. Sci.* **26** (12), 1777-1784.

Buck, B. C. (1956). *Advanced calculus*, McGraw-Hill, New York.

Burrows, W. J. (1992). *Weather cycles—real or imaginary?*, Cambridge UP, Cambridge.

Cairns-Smith, A. G. (1982). *Genetic takeover and the mineral origin of life*, Cambridge UP, Cambridge.

Cao, J.-Y. and Cao, B.-G. (2006). Design of fractional order controllers based on particle swarm optimization.

The Capitol Net, Smart Gird, Government series, Alexandria 2009.

Carbone, A. and Gromov, M. (2001). *A mathematical slices of molecular biology*, Supplement to volume 88 of Gazette des Mathématiciens, French Math. Soc. (SMF), Paris.

Chakraborty, K., Kanemitsu, S., and Tsukada, H. (2009). *Vistas of special functions II*, World Sci., New Jersey-London-Singapore etc.

Chakraborty, K., Kanemitsu, S., and Kuzumaki, T. (2016). *A quick introduction to complex analysis*, World Sci., Singapore etc.

Chen, S., Dzeng, R. S., Yang, M., Chlu, K., Shlen, G. and Wal, M. C. (1994). Arsenic species in ground waters of the black foot disease area Taiwan, *Taiwan. Environ. Sci. Technol.*, **28**, 877-881.

Chen, W., Kanemitsu, S., and Sato, K. (2008). A classification of amino acids in terms of their specificities-polarity, charge and codons, *Kayanomori*, **8**, 9-12.

Clavier, J. M. and Notredame, C. (2007) *Bioinformatics for dummies*, Wiley, Indianapolis.

De (1999). *Committee of the driving system of EVs, Electrical Engineering Society, The latest technologies of EVs*, Ohm-sha, Tokyo.

Dansgaard, W. *et al.* (1993). Evidence for general instability of past climate from a 250-kyr ice-core record, *Nature* **364**, (6434), 218-220.

Dettman, J. B. (1984). *Applied complex variables*, Dover, New York (Reprint, first published by Macmillan in 1965).

Dirac, P. (1958). *The principles of quantum mechanics*, 4th ed. Oxford UP, Oxford.

Donoghue Jr., W. J. (1969). *Distributions and Fourier transforms*, Academic Press, New York-London.

Ebara, A. (1987). *The origin and evolution of humanity*, NHK Books, Toyko.

Erdélyi, A., Magnus, W., Oberhettinger, F. and Tricomi, F. G. (1953). *Higher Transcendental Functions*, Vols. I-III, Based, in part, on notes left by Harry Bateman, McGraw-Hill, New York.

EC Confidential LLC (2006). *Who killed the electric car?*

Newest technology for EVs, Soc. Electr. Engrg, Res. Committee for the driving systems for EVs, Ohmsha, Tokyo 1999.

Falb, P. (1990). *Methods of algebraic geometry in control theory: Part I Scalar linear systems and affine algebraic geometry*, Birkhauser Basel.

Falb, P. (2000). *Methods of algebraic geometry in control theory: Part II Multivariable linear systems and projective algebraic geometry*, Birkhauser Basel.

Feiler, C. and Schreich, W. P. (2013). Entanglement and analytical continuation: an intimate relation told by the Riemann zeta function, *New J. Phys.* **15**, 063009.

Feiler, C. (2013). *Quantum physics and number theory connected by the Riemann zeta function*, Thesis Univ. Ulm.

Feng, X., Ziurntos, T. and Butler-Purry, K. L. (2010). Dynamic load management for NG IPS ships, *Proc. IEEE Power Engineering Society General Meeting*, Minneapolis.

Fukuoka, S. (2009). *Between the living and the non-living*, Kodansha, Tokyo (in Japanese).

Fujita, K. (1997). *Let's become as healthy as cavemen*, Shin-cho-sha, Tokyo.

Furuhama, S. (1970). *Engineering of inner combustion engines*, Sangyou-tosho, Tokyo (in Japanese).

Furukawa, O. (2007). Physics through cars, Ohmsha, Tokyo.

An adaptation method for removing arsenate species from water solution, Pure Appl. Math. J. 4 (2-1), 47-54, doi: 10.11648/j.pamj.s.2015040201.19

Garai, R., Chakraborty, A. K., Dey, S. B. *et al.* (1984). Chronic arsenic poisoning from tube-well water, *J. Indian Med. Assoc.* **82**, 34-35.

Gimarc, B. M. (1979). The shapes of simple polyatomic molecules and ions. I. The series HAAH and BAAB, *J. Amer. Chem. Soc.* **92**, 266-275.

Giardina, Ch. R. Band-limited signal extrapolation by truncated Bernstein polynomials, *J. Math. Anal. Appl.* **104**, 264-273.

GoldBerg, S. and Glanbig, R. A. (1988). Anion sorption on a calcareous, montmorillonitic soil-arsenic, *Soil Sci. Am. J.* **52**, 1297-1300.

Granger, F. S. (1985). *Fluid mechanics*, Dover, New York.

Grodins, F. S. (1963). *Control theory and biological systems*, Columbia Univ. Press, New York and London.

Grunsky, H. (1983). *The general Stokes' theorem*, Pitman, Boston-London-Melbourne.

Hardy G.-H. and Riesz, M. *The general theory of Dirichlet's series*, CUP, Cambridge; reprint, Hafner, New York 1972.

Hayakawa, H. *The mineral water guidebook*, Shin-cho-sha, Tokyo, 1995.

Hayakawa, Y. (2008). *Systems and their controlling*, Ohmsha, Tokyo (in Japanese).

Helton, J. W. and Merino, O. (1998). *Classical control using H^∞ methods, Theory, optimization and design*, SIAM, Philadelphia.

Hiramatsu, S. and Saito, S. (2016). *An Introduction to non-Abelian Class Field Theory*, World Sci.

Hirayama, N. (2014) *The body functions viewed at the molecular level, Kodan-sha*, Tokyo 1998. Singapore etc.

Hossain, H., Han, Z. and Poor, H. V. (2014). *Smart Grid communications and networking*, Cambridge UP, Cambridge.

Huxley, M. N. (1972). *The distribution of prime numbers—Large sieves and zero density theorems*, Oxford UP, Oxford.

Ignalo, L. (2007). *Say yes to NO program*, Nikkei BK Publ. Center, Tokyo.

Ikeda, Y. (1993). *Introduction to bio-material*, Gakkai-shuppan Center, Tokyo.

Imai, I. (1963). *Fluid mechanics I, II*, Science-sha, Tokyo (in Japanese).

Ivić, A. (1985). *The Riemann zeta-function*, Wiley, New York.

Kakita, T. (1985). *Introduction to the theory of Schwarz distribution*, Nihon-hyoron-sha, Tokyo.

Kanemitsu, S., Li, F.-H., Kamata, Y. and Maruyama, K. (2016). Mathematical foundations of fluid dynamics, *Basic Sciences J. Textile Univ.* **29**, 275-285.

Kanemitsu, S., Li, F.-H. and Sato, K. (2010). Working principles of an inner combustion engine, *Kayanomori* **13**, 41-48.

Kanemitsu, S. and Tsukada, H. (2007). Vistas of special functions, World Sci., New Jersey-London-Singapore etc.

Kanemitsu, S. and Tsukada, H. *Contributions to the theory of zeta-functions—modular relation supremacy*, World Sci., Singapore etc.

Kanemitsu, S., Li, F.-H., Kamata, Y. and Maruyama, K. (2016). Mathematical foundations of fluid dynamics, *Basic Sciences Journal of Textile Universities*, **29** No. 3, 275-290.

Kawashima, S. (1982). *To be healthy by alkali meals*, Shincho-sha, Tokyo.

Kennedy, J. and Eberhart, R. C. (1995). *Particle swarm optimization*, Proc. IEEE Intern. Conf. Neural Networks, 1942-1948, Piscateway, New Jersey.

Kimura, H. (1997). *Chain scattering approach to H^∞-control*, Birkhäuser, Boston-Basel-Berlin.

Kitano, S. (1972). *Introduction to inner combustion engines*, Nikkan Kogyou Shinbun-sha Tokyo.

Khono, A. (1989). *Science of frictions*, Shokabo, Tokyo.

Koide, S. (1975). *Physics*, Shokabo, Tokyo.

Koide, S. (1971). *Quantum theory*, 6th ed. Shokabo, Tokyo.

Kondo, H. *Basic study on the removal of harmful metalic objects including arsenate from polluted undergroond water*, Doctoral thesis, Kyushu Univ.

Kondo, H. (1997). An outline of arsenic pollution of ground waters in the southern area of Fukuoka Prefecture, *Japan Soc. Water Environ.* **20** (7), 438-442.

Kondo, S. (1993). *Health effects of low-level radiation*, Medical Physics Publ., Madison.

Kumagai, N. Ciphers in the times of the Web–Challenge of net security, Chikuma-shobo, Tokyo.

Kundurm, D., Mashayehk, S., Ziurntos, T. and Butler-Purry, K. L. (2014). *Cyber-attack impact analysis of smart grid, in H. Hossain, Z. Han and H. V. Poor, Smart Grid communications and networking*, Cambridge UP, Cambridge, 354-372.

Lamb, H. (1945). *Hydrodynamics*, Reprint Dover, New York.

Latorre, J. I. and Sierra, J. (2015). There is entanglement in primes, *Quantum information & computation archive* **15**, Issue 7-8, 622-659.

Li, F.-H., Galhotra, S. and Kanemitsu, S. (2015). Emerging importance of EVs in the Green Grid Era, *Pure Appl. Math. J. Special issue: Mathematical aspects of engineering disciplines* (ed. by H.-L. Li), Volume **4**, Issue 5-1, 38-45 DOI: 10.11648/j.pamj.s.2015040501.18.

Li, F.-H., Kamata, Y., Kanemitsu, S. and Maruyama, K. (2018a). EVs in Green Grid Era and climate change, *Comm. Num. Anal.*, Available online at www.ispacs.com/cna Volume **2018**, Issue 1, Year, Article ID cna-00330, 20 Pages doi:10.5899/2018/cna-00330.

Li, F.-H., Kamata, Y., Kanemitsu, S. and Maruyama, K. (2018b). Vectorial version of the general Stokes theorem and the Karman vortex street, *Comm. Num. Anal.*, Available online at www.ispacs.com/cna Volume **2018**, Issue 1, Year 2018 Article ID cna-00331, 15 Pages doi:10.5899/2018/cna-00331.

Li, F.-H. and Kanemitsu, S. (2919). *Around boundary functions of the right half-plane and the unit disc*, Advances in Special functions and analysis of Differential Equations, to appear.

Li, F.-H., Kanemitsu, S. and Sato, K. (2010). Entropy and heat efficiency of the inner combustion engine, *J. Math. Sci.; Adv. & Appl.* **6**, No. 1, 99-111.

Li, F.-H., Wang, N.-L. and Kanemitsu, S. (2012). *Number theory and its applications*, 194 pages, World Sci., London-Singapore-New Jersey, Nov. 28.

Li, F.-H., Kanemitsu, S. and Li, H.-Y. Global warming and EV2GG system—ants trying to fight an elephant, Diandong qiche xinjishu yanjiu (Research on new techniques for EVs) (F. H. Li and S. Kanemitsu editors), 171 pages, Shaanxi Sci. Techn Press, Xi'an, 2017, 15-31

Li, F.-H. (2019). Robust controller for servo systems, to appear.

Li, H.-L., Li, F.-H., Wang, N.-L. and Kanemitsu, S. (2017). *Number theory and its applications II*, 285 pages, World Sci., London-Singapore-New Jersey.

Li, H.-L. and Li, F.-H. (2019). PSO optimization for cell-balancing charge of Lithium ion batteries, to appear.

Lichtenstein, L. (1929). *Grundlagen der Hydromechanik*, Springer, Berlin.

Lighthill, M. J. (1958). *Generalized functions*, CUP, London.

Longman (1978). *dictionary of contemporary English*, Essex.

Mack, R., Dahl, J. P., Moya-Cessa, H., Stunz, W. T., Wlser, R. and Schreich, W. P. (2010). Riemann ζ from wavepacket dynamics, *Phys. Rev. A* **87**, 032119.

Merzbacher, E. (1961). *Quantum mechanics*, 2nd ed., John Wiley, New York.

Miyamoto, T. (2015). *Acceleration switch on—Making EV yourself*, CQ Publishing, Tokyo.

Mizohata, S. (1965). *The theory of partial differential equations*, Iwanami-shoten, Tokyo.

Moon, D. H., Kim, K. W. and Yoon, I. H. *et al.* (2011). Stabilization of arsenic-contaminated mine tailings using natural and calcined oyster shells, *Environ. Earth Sci.* **64** (9), 597-605.

Murayama, T. and Tsunemoto, H. (2008). *Engineering of automobile engine*, Tokyo Denki University Press Tokyo.

Murota, T. (1984). *Developments of entropy economy*, Hirokawa-shoten, Tokyo, 187-202.

Nakahara, M. and Ohimi, T. (2008). *Quantum computation–From linear algebra to physical realizations*, CRC Press, Boca Raton etc.

Nasu, H. and Higasa, H. (1995). Development of yonden electric vehicle "PIVOT", *JSAE Review* **16**, 77-82.

Neunzert, H. (1985). The road vehicle system and related mathematics, *Proc. Workshop*, March 18-22, 1985, Teubner, Stuttgart.

Nielsen, M. A. and Chuang, I. L. (2010). *Quantum Computation and Quantum Information*, Cambridge UP, Cambridge.

Nishioka, H. (1948). *History of cold and hot—a theory of periodic climate change in 700 years*, Kogakusha, Tokyo.

Nishioka, H., Uchida, T., Uchida, K., Yazawa, T. and Amako, T. (2011). Optimization of conditions for heat treatment of oyster shells which improves the adsorption rate of ions that give load on environment, *J. Soc. Industr. Wastes, Resources Recycling* **22**, 276-283.

Nishiyama, T., Endoo, K. and Wada, A. (2016). *In-wheel motors—Principles and design*, Kagaku-joho-shuppan, Tokyo.

NPO Japan Supplement Organization (2008). *Supplements bible 2008-2009*, Shogakukan, Tokyo (in Japanese).

Okamura, M. (1999). *Electric double layer capacitor and energy capacitor system (ECS)*, Nikkan-kogyo-Shinbun-sha, Tokyo.

Okamoto, S. (1995). *Fluid mechanics*, Morikita-shuppan, Tokyo.

Okamura, M. (1999). *Electric double layer capacitor and energy capacitor system (ECS)*, 1 Nikkan-kogyo-Shinbun-sha, Tokyo.

Paine, C. (2011). (ed.), Revenge of the electric car.

Pauling, L. and Wilson, E. B. (1935). *Introduction to quantum mechanics*, McGraw-Hill, New York.

Papoulis, A. (1962). *The Fourier integral and its applications*, McGraw-Hill, New York.

Pedlovsky, J. (1987). *Geophysical fluid dynamics*, 2nd ed., Springer, New York etc.

Peirce, J. J., Weiner, R. F. and Wesilind, P. (1998). *Environmental pollution and control, Butterworth-Heinemann*, Woburn (Chapter 16).

Pinkus, G. and Sternlicht, B. (1961). *Theory of hydrodynamic lubrication*, McGraw-Hill, New York.

Pohl, F., Chernobyl, (1987). Bantam Books, Toronto etc.

Pollack, R. (1999). *The missing moment: How the unconscious shapes modern science*, Houghton Mifflin Harcourt, New York.

Li, Y., Qin, Z. and He, X.-S. (2004). Supervisor-student model in particle swarm optimization, in *Proc. IEE Congr. Evol. Comput.*, Piscataway, USA, 542-547.

Qiu, R. C. and Antonik, P. (2014). *Smart Grid and Big Data: Theory and Practice*, Wiley, New York.

Rentrop, P. *Numerische Porbleme in der Fahrezeugsdynamik*, [Neunzert (1985)], 44-58.

Rozenfel'd, A. S. and Yahinson, B. I. (1966). Transitional processes and generalized functions, Nauka, Moscow.

Ruderman, D. L. and Bialek, W. (1992). *Seeing beyond the Nyquist limit*, Neural Computation 4(5):682-690.

Rudin, W. (1986). *Real and complex analysis*, MacGraw-Hill, New York etc.

Sakurai, J. (2011). *Which parts of nuclear plants are dangerous?–Accidents in the world and Fukushima*, Asahi Shinbun Publ., Tokyo.

Sato, M., Yamamoto, G., Imura, T. and Fujimoto, H. (2014). *Efficiency evaluation in wireless power transfer via magnetic resonance coupling to in-wheel motor*, IEEJ-MD14094.

Schey, H. M. (1992). *Div, grad, curl, and all that*, W.W-Norton & Co. New York.

Sena, H. and Ohta, S. (2007). *Power of mitochondria*, Shincho-sha, Tokyo (in Japanese).

Serre, J.-P. (1973). *A course in arithmetic*, Springer Verl., New York-Heidelberg-Berlin.

Seymour, J. and Girardet, H. (1988). *Blueprint for a green planet*, Dorling Kindersley Publ. London.

Shapiro, A. H. (1953). *The dynamics and thermodynamics of compressible fluids.*

Splettstösser, W. (1985). *Some aspects on the reconstruction of sampled signal functions, the Road-Vehicle-System and Related Mathematics, H. Neunzert (ed.), B. G. Teubner*, 126-142.

Stewartson, K. (1964). *The theory of laminar boundary layers in compressible fluid flows*, OUP London.
Stimmel, C. L. (2015). *Big Data Analytics Strategies for the Smart Grid*, CRC Press, New York.
Sumi, A. (1999). *The truth of global warming*, Wedge, Tokyo.
Suzuki, K., Sogabe, K. and Shimosaka, H. (1984). *Dynamics of machinery*, Jikkyoshuppan, Tokyo.
Tada, R. (2013). *Scientific study on climate change*, Misuzu-shobo, Tokyo.
Takagi, N. and Watanabe, M. (2011). *Radiation on the dining table*, Nanatsumori Shokan, Tokyo.
Takeda, K. (2011). *The Nuke accident–danger of radiation by remains*, Asahi Shinbun-sha, Tokyo.
Takeda, K. (2011). *How can you protect your kids from radiation?*, Shufu-to-seikatsu-sha, Tokyo.
Tamagawa-seiki co. (2002). (ed.), *Introduction to effective use of gyro technology*, Kogyo-chosakai, Tokyo.
Tanaka, Y. (2010). *Let's buy an EV at Yamada electricity shop*, Random House Japan, Tokyo.
Tatuzawa, T. (1980). *Theory of functions*, Kyoritsu-shuppan, Tokyo.
Terou, S. and Majima, T. (1999). *Free radical–its roles ranging from life and environement to the frontier technology*, Yoneda Shuppan, Ichikawa.
Tipler, F. (1994). *The physics of immortality*, Pan Books, London.
Titchmarsh, E. C. (1986). *The Theory of the Riemann zeta-function*, 2nd ed., Clarendon Press, Oxford.
Tominaga, K. (2011). *How should we eat to protect ourselves from radiation*, Kawadeshobo-shinsha, Tokyo.
Torosov, B. T., Della Valle, G. and Longhi, S., Quantum simulation of the Riemann ζ function, Phys. Rev. A **87** (2013) 032103.
Toyokura, T. and Kamemoto, K. (1976). *Fluid mechanics*, Jikkyo-shuppan, Tokyo.
Tsukamoto, K. (2010). *EV wars*, Asahi Publ. House, Tokyo.
Tsunematsu, T., Uematsu, E., Saito, K. and Tamura, H. (2012). Anti-dissolvation of naturally existing arsenate species by addition of gypsum powder and its mechanism, *Rep. Agr. Rural Industr.* **80**, 141-150.
Wada, E. *et al.* (1995). *Stable isotopes in the biosphere*, Kyoto Univ. Press Kyoto.
Wang, Sh.-Ch. and Liu, Y.-H. (2018). PSO-based Fuzzy logic optimization of dual performance characteristic indices for fast charging of Lithium-ion batteries, *Energies*, **11**, 469; doi:10.3390/en11020469.
Weaver, J. A. (1983). *Applications of discrete and continuous Fourier transforms*, Wiley, New York etc.
Weil, A. (1995). *Spontaneous healing*, Ballantine Book New York. Translation into Japanese, Kadokawa-Shoten, Tokyo.
Whittaker, E. T. (1915). *On the functions which are represented by the expansions of the interploation theory*, Proc. Royal Soc. Edinburgh **35**, 181-194.
Widder, D. W. (1929). A generalization of Dirichlet's series and of Laplace's integrals by means of a Stieltjes integral. Trans. Amer. Math. Soc. **31**, 694-743.

Widder, D. W. (1946). The Laplace transform, Princeton UP, New Jersey.
Woodward, P. M. (1953). *Probability and information theory, with applications to radar*, Pergamon Press, London.
Widder, D. W. (1956). *Advanced calculus*, Prentice-Hall, Reprint Dover New York.
Wintner, A. (1925). *Spektraltheorie der unendlichen Matrizen—Einführung in den analytischen Apparat der Quantenmechanik*, Hirzel, Leipzig.
Xin, X. and Kimura, H. (1994). $(J, J^p rime)$-lossless factorization for strictly proper functions, *Int. J. Control* **205-206**, 1289-1318.
Xin, X. and Kimura, H. (1994). Singular $(J, J^p rime)$-lossless factorization for descriptor systems, *Linear Alg. its Appl.* **59**, 1383-1400.
Yamamoto, T. *History of Japan as told by climate*, Societe, Tokyo.
Ye, M., Bai, Z.-F. and Cao, B.-G. Robust H2/Hinfinity control for regenerative braking of electric vehicles, ICCA 2007. *IEEE International Conference on Control and Automation*, 1366-1370.
Yoshida, K. (1956). *Modern analysis*, Kyoritsu Shuppan, Tokyo (in Japanese).
Yoshida, M. (1995). *How we are to understand the molecular orbital method*, 2nd ed., Tokyo Kagaku-dojin, Tokyo (in Japanese).
Zong, Q.-Z. and Hornik, T. (2014). *Control of power inverters in renewable energy and smart grid integration*, Wiley, New York.

Index

www.ingramcontent.com/pod-product-compliance
Ingram Content Group UK Ltd.
Pitfield, Milton Keynes, MK11 3LW, UK
UKHW022103170926
8624IPUK00024B/47

9 789811 206962